Ergebnisse der Mathematik und ihrer Grenzgebiete

Band 13

Herausgegeben von
P. R. Halmos · P. J. Hilton · R. Remmert · B. Szőkefalvi-Nagy

Unter Mitwirkung von
L.V. Ahlfors · R. Baer · F. L. Bauer · R. Courant
A. Dold · J. L. Doob · S. Eilenberg · M. Kneser · G. H. Müller
M. M. Postnikov · B. Segre · E. Sperner

Geschäftsführender Herausgeber: P. J. Hilton

Beniamino Segre

Some Properties of Differentiable Varieties and Transformations

With Special Reference to the Analytic and Algebraic Cases

Second Edition

With an additional part written in collaboration with J. W. P. Hirschfeld

Springer-Verlag Berlin Heidelberg New York 1971

Prof. BENIAMINO SEGRE

Università di Roma, Istituto Matematico
,,Guido Castelnuovo", I-00100 Roma

AMS Subject Classifications (1970):
14-02, 14 E 05, 34 A 30, 35 A 30, 35 G 05, 35 K 10, 50-02, 50 DXX, 53-02, 53 A 20

ISBN-13:978-3-642-65008-6 e-ISBN-13:978-3-642-65006-2
DOI: 10.1007/978-3-642-65006-2

Library of Congress Catalog Card Number 72—137498.
Softcover reprint of the hardcover 2nd edition 1971

Beniamino Segre

Some Properties of Differentiable Varieties and Transformations

With Special Reference
to the Analytic and Algebraic Cases

Second Edition

With an additional part
written in collaboration with
J. W. P. Hirschfeld

Springer-Verlag New York Heidelberg Berlin 1971

Prof. Beniamino Segre
Università di Roma, Istituto Matematico
„Guido Castelnuovo", I-00100 Roma

AMS Subject Classifications (1970):
14-02, 14 E 05, 34 A 30, 35 A 30, 35 G 05, 35 K 10, 50-02, 50 DXX, 53-02, 53 A 20

ISBN-13:978-3-642-65008-6 e-ISBN-13:978-3-642-65006-2
DOI: 10.1007/978-3-642-65006-2

Library of Congress Catalog Card Number 72—137498.
Softcover reprint of the hardcover 2nd edition 1971

Preface to the First Edition

The present volume contains, together with numerous addition and extensions, the course of lectures which I gave at Pavia (26 September till 5 October 1955) by invitation of the «Centro Internazionale Matematico Estivo». The treatment has the character of a monograph, and presents various novel features, both in form and in substance; these are indicated in the notes which will be found at the beginning and end of each chapter.

Of the nine parts into which the work is divided, the first four are essentially differential in character, the next three deal with algebraic geometry, while the last two are concerned with certain aspects of the theory of differential equations and of correspondences between topological varieties. A glance at the index will suffice to give a more exact idea of the range and variety of the contents, whose chief characteristic is that of establishing suggestive and sometimes unforeseen relations between apparently diverse subjects (e. g. differential geometry in the small and also in the large, algebraic geometry, function theory, topology, etc.); prominence is given throughout to the geometrical viewpoint, and tedious calculations are as far as possible avoided.

The exposition has been planned so that it can be followed without much difficulty even by readers who have no special knowledge of the subjects treated. Although, for reasons of space, various proofs are here only sketched, nevertheless the essence of the methods is always made clear, while, on the other hand, at the end of each chapter there are sufficient references to enable the reader to pursue matters further. However, in certain cases cognate results have been omitted for the sake of brevity; for the same reason, the Bibliography does not claim to be complete.

I have had the good fortune to have received valuable assistance from Dr. David Kirby in preparing the English version of the text, from Dr. Edoardo Vesentini in typing and revising the typescript, and from Prof. Leonard Roth in reading the proofs; to all these I tender hearty and grateful thanks. In conclusion, I wish to express to the Springer-Verlag my appreciation of the superlatively good printing and production of this book.

Roma, 22. II. 1957 Beniamino Segre

Preface to the Second Edition

This edition differs from the first mainly in the insertion of a new Part Nine entitled "Projective Differential Geometry of Systems of Linear Partial Differential Equations", thus filling a gap in the theory of differentiable varieties. Part Nine of the previous edition is now Part Ten. Various misprints and errors have also been corrected. Otherwise the first edition remains intact.

I am most grateful to Dr. JAMES HIRSCHFELD for his valuable assistance in the compilation of the new Part and for his assiduity in tracing the misprints.

Roma, 29. IX. 1970 BENIAMINO SEGRE

Contents

Part One

Differential Invariants of Point and Dual Transformations

In this first Part we shall show that we can determine a *complete system of differential invariants of the first order*, relative to a pair of differential elements homologous in a point or dual correspondence, between portions of two Euclidean spaces, which is biregular and of class C^1. From such metric invariants will be deduced certain topological invariants relative to the united points of correspondences between superimposed varieties, as well as some projective invariants belonging to a pair of elements common to two dual correspondences, and also to two hypersurfaces of a hyperspace which touch at a common point. A deeper study of the above invariants will appear in the following two Parts.

§ 1. Local metrical study of point transformations. — Let E_n, E'_n be two oriented (real) Euclidean spaces of dimension n ($\geqq 1$), and let T be any biregular correspondence of class C^1 between two regions therein. If T transforms the point $P(x_1, x_2, \ldots, x_n)$ of E_n into the point $P'(X_1, X_2, \ldots, X_n)$ of E'_n, then the equations of T express $X_1, X_2, \ldots, X_n$ as functions of $x_1, x_2, \ldots, x_n$, and the associated Jacobian matrix

$$J = \frac{\partial(X_1, X_2, \ldots, X_n)}{\partial(x_1, x_2, \ldots, x_n)}$$

has a non-zero determinant.

In order to obtain the first order differential invariants of T for the pair (P, P'), we observe that, if a point Q of E_n — near to P — tends to P in such a way that the line (semi-line) PQ tends to a line (semi-line) r through P, but otherwise arbitrarily, then the homologue $Q' = T(Q)$ tends to $P' = T(P)$ in such a way that the line (semi-line) $P'Q'$ assumes a limiting position r' depending only on r. It is well known, and can be immediately verified, that the correspondence defined in this manner between the lines (semi-lines) r and r' produces a non-degenerate homography between the stars with centres P and P'.

We have, moreover, that — when $Q \to P$ — the quotient of the lengths of the segment $P'Q'$ and PQ assumes a (positive) limit which depends only on the direction of r; this number and its reciprocal are called, respectively, the *coefficient of dilatation* and the *coefficient of contraction* of T at P in the direction of r.

The required invariants are deduced from the study of the way in which the above coefficients depend on the corresponding directions. For this purpose, on the different semi-lines of E_n through P, we take a segment equal to the associated coefficient of contraction; we then have that the free extremity of this segment generates a hyperellipsoid $\mathfrak{J}$ with centre P, called the *hyperellipsoid of deformation* of T belonging to P.

It is easily seen that (to within infinitesimals) $\mathfrak{J}$ is similar to the transform by T^{-1} of an infinitesimal hypersphere of E'_n with centre P', and that the above projectivity, which is subordinate to T, between the stars with centres P and P' transforms two conjugate diameters of $\mathfrak{J}$ into two lines of E'_n through P' which are perpendicular. Hence there exist n lines through P, any two being perpendicular, which are transformed into lines through P' any two of which are perpendicular; such lines — called the *principal lines* of T through P — are evidently not determined, if, and only if, $\mathfrak{J}$ is a hyperellipsoid of rotation, and they can be characterized as those lines through P for which the corresponding coefficient of contraction, or dilatation, has an extremum (not, however, necessarily a maximum or a minimum), wherefore to each such coefficient we give the further attribute of *principal.* It is clear that the n *principal coefficients of contraction* of T at P give the lengths of the semi-axes of $\mathfrak{J}$; and, moreover, it is evident which modifications occur in the foregoing when $\mathfrak{J}$ is a hyperellipsoid of rotation.

Analogous entities are obtained in E'_n, associated with the correspondence T^{-1}. It is evident that the principal coefficients of contraction of T at P are equal to the principal coefficients of dilatation of T^{-1} at P', and that the two principal n-ads with vertices P and P' correspond in the homography induced by T between the stars of semi-lines through P and P'; this last associates with a positive orientation of the first n-ad an orientation of the second, which is positive or negative according as J has a positive or negative determinant. We now observe that:

A point transformation T, which is reversible and of class C^1 between two Euclidean spaces of dimension n, possesses n, and only n, metrical differential invariants of the first order which are mutually independent. More precisely, the n principal coefficients of contraction of T are such a set of n invariants, and any other invariant of the first order is a function of them.

The invariance of those coefficients is implied by their definition. The other properties asserted follow from:

There exists, between E_n and E'_n, one, and only one, affine transformation, A, approximating to T in the neighbourhood of the first order for the pair (P, P'), and it satisfies the following conditions:

1. A transforms P to P' and, between the stars of semi-lines with centres at these points, gives the same homography as that subordinate to T;

2. The coefficients of contraction of A and T at P are equal in every direction.

If Cartesian coordinates are introduced in E_n and E'_n referred to the two principal n-ads belonging to P and P', this may require a change in the positive orientation in one of the two spaces; then the equations of A assume the following simple form:

$$X_1 = c_1 x_1,\ X_2 = c_2 x_2, \ldots, X_n = c_n x_n,$$

where $c_1, c_2, \ldots, c_n$ denote the n principal coefficients of dilatation. It is clear that *the affine transformation A transforms the hyperellipsoid of deformation $\mathfrak{J}$, belonging to P, into the hypersphere of E'_n with centre P' and radius* 1; A can therefore be characterized uniquely by this property together with that of transforming, one into the other, the two principal n-ads associated with P and P', so determining convenient relations between them and their orientation. It is now immediately seen that the product of the n principal coefficients of dilatation is equal to the absolute value of the determinant of J; this we may call the *density* of the transformation T belonging to the pair (P, P'), and it appears as the ratio of the volumes of two homologous infinitesimal regions of E_n and E'_n containing P and P' respectively. This result is included in the following theorem, giving explicit expressions for n differential invariants of the first order for T, which are generally mutually independent by the foregoing.

The sum of the squares of the products of the principal coefficients of dilatation taken t at a time $(1 \leq t \leq n)$, is equal to the sum of the squares of the $\binom{n}{t}^2$ minors of order t extracted from the matrix J.

§ 2. Some topologico-differential invariants. — In view of its local character, the preceding development is immediately seen to be capable of extension to point correspondences between Riemannian varieties. Again, therefore, in this case we can define *the principal tangent lines*, the *principal coefficients of contraction*, etc., which suggests the introduction, and the study, of curves which may be called *principal*, namely of curves having at any point a principal line as tangent. Without dwelling on it at present, we notice that a correspondence is *conformal* if, and only if, the principal lines are wholly indeterminate. Of particular interest will be the case of partial indetermination, characterized by the property that at any point the hyperellipsoid of deformation is a hyperellipsoid of rotation (in one of the several possible ways).

We now refer in particular to the case in which (with the notation of § 1, p. 1) we have $E_n = E'_n$, $n = 1$ and $P \equiv P'$, so that T is a correspondence on a line having a united point P. With such hypotheses, to the coefficient of dilatation of T at P can be attributed a sign, taking the positive sign or negative sign according as, in the neighbourhood of P, T is concordant or discordant. Hence the above coefficient is, in either case, equal to the value assumed by the derivative $\frac{dX}{dx}$ at P, where

x and X denote the coordinates of two corresponding points of the superimposed lines E_1, E_1', with the same reference system. It is seen immediately that this expression remains unchanged by any alteration to the system of coordinates, that is by replacing x with any function of the same x of class C^1, having non-zero derivative at P, and consequently replacing X with that same function of X; this result holds manifestly not only in the field of real numbers, but likewise in the field of complex numbers. The aforesaid coefficient of dilatation therefore has a *topologico-differential* significance, in so far as it depends only on T and P, and not on the choice of coordinates in E_1. This allows us to substitute for E_1, in the foregoing, any curve that can be considered as arising from E_1 by means of a transformation of class C^1. Finally this result can be extended to the case of arbitrary n, in the manner which we now proceed to indicate.

Let P be a simple point of a variety V_n, which is differentiable (or complex) in a neighbourhood of P, and let S_n denote the real (or complex) projective tangent space to V_n at P. Let us then consider a correspondence T of V_n into itself for which P is a united point and which, in a neighbourhood of P, is reversible and differentiable (or analytic). In the star of ∞^{n-1} lines tangent to V_n at P, or rather the star of real (or complex) lines of S_n passing through P, T defines intrinsically, as has been noted (§ 1, p. 1), a non-degenerate homography — which we shall shortly call the *tangent homography* of T at P — possessing generally n united lines which are distinct. The coefficients of dilatation of T in the separate directions of those lines provide, then, n numbers (generally complex) which are the *topologico-differential invariants* of T at P; we shall call them simply the *coefficients of dilatation of T at P*, and we shall see later (§ 3, p. 6) that with each of them can be associated a certain cross-ratio having a geometric significance.

It follows that, in order to calculate the invariants noted above, we can introduce in V_n any allowable coordinate system. If T transforms the point $(x_1, x_2, \ldots, x_n)$ into the point $(X_1, X_2, \ldots, X_n)$, and if J denotes the Jacobian matrix of $X_1, X_2, \ldots, X_n$ with respect to $x_1, x_2, \ldots, x_n$ evaluated at the point P, then *the above cross-ratios are none other than the roots of the characteristic equation of J*. Their product is then precisely equal to the determinant of J, which is therefore also a topologico-differential invariant of T at P; this is also verifiable immediately in a direct manner.

More generally, let V_n and V_n' be two differentiable (or complex) varieties, between which there are two differentiable (or complex) correspondences T_1 and T_2, both being reversible in a neighbourhood of one of their common pairs of homologous points (P, P'). Then we can apply the foregoing to the transformation $T = T_1 T_2^{-1}$, which is a self-transformation of V_n leaving P fixed; this supplies n *topologico-differential invariants* *of T_1 and T_2 relative to a common pair (P, P').* From what has been said above and in § 1, p. 3 we have immediately, in particular, that:

The ratio of the densities of T_1 and T_2, belonging to the common pair (P, P'), constitutes a topologico-differential invariant relative to it (it is equal to the density of T at its united point P).

§ 3. Projective construction of the above invariants. — With the object of arriving rapidly at the results which were alluded to in § 2, p. 4, let us consider a projective space (real or complex) S_{2n-1}, with odd dimension $2n-1$. Three subspaces S_{n-1}, each pair of which are skew, are without projective invariants; the equations of such spaces can always be reduced to the form

$$S^{(1)}_{n-1}: x_1 = x_2 = \cdots = x_n = 0, \qquad S^{(2)}_{n-1}: X_1 = X_2 = \ldots = X_n = 0,$$
$$S^{(3)}_{n-1}: x_1 = X_1,\ x_2 = X_2, \cdots, x_n = X_n.$$

Then we are still allowed a change of coordinates such that these equations are preserved; this — as is immediately seen — is obtained, most generally, by operating with the same arbitrary reversible linear substitution on $x_1, x_2, \ldots, x_n$ and on $X_1, X_2, \ldots, X_n$. There are ∞^{n-1} lines meeting the three spaces; these lie on a V^n_n of C. SEGRE, and its equations are

$$\varrho x_i = u_i, \quad \varrho X_i = \lambda u_i \qquad (i = 1, 2, \ldots, n),$$

where ϱ is merely a factor of proportionality of the coordinates, $u_1, u_2, \ldots, u_n$ are n homogeneous parameters which have constant ratio for any one of those lines, and λ is a non-homogeneous parameter with the significance of a cross-ratio of a point of V^n_n together with the three points of the given spaces on the line of V^n_n containing that point (the points of the given spaces are obtained by $\lambda = \infty$, 0, 1 respectively).

Now let us consider in S_{2n-1} a fourth sub-space $S^{(4)}_{n-1}$, which meets neither $S^{(1)}_{n-1}$ nor $S^{(2)}_{n-1}$; its equations can be written in the form

$$X_i = \sum_{j=1}^{n} a_{ij} x_j \qquad (i = 1, \ldots, n),$$

where the matrix A of the a_{ij}'s has determinant different from zero. Such an $S^{(4)}_{n-1}$ — if it is generally situated — meets the above V^n_n in n distinct points; and it is clear that *the values assumed by the parameter λ at such points are none other than the latent roots of the matrix A.*

Removing the arbitrariness that still exists in the choice of coordinates and extending S_{2n-1}, if need be, to the complex field, we can, in general, reduce the equation of $S^{(4)}_{n-1}$ to the form

$$X_i = a_i x_i \qquad (i = 1, 2, \ldots, n),$$

where $a_1, a_2, \ldots, a_n$ are the latent roots referred to above. We have then:

Four S_{n-1}'s of S_{2n-1} admit in general n independent projective invariants and no more. Such invariants can be taken as the cross-ratios of the sets of four points cut by the four S_{n-1}'s on the n lines of S_{2n-1} which meet them.

Having commenced with this result, we now refer back to the two correspondences T_1 and T_2 of the penultimate paragraph of § 2, p. 4. On the product variety $W_{2n} = V_n \times V'_n$ they are represented by two n-dimensional varieties, $M_n^{(1)}$ and $M_n^{(2)}$, both passing simply through the point $O = P \times P'$, such that their tangent spaces have a line in common neither with each other, nor with that of the variety $L_n^{(1)} = V_n \times P'$, nor with that of $L_n^{(2)} = P \times V'_n$. Let

$$S_{2n-1}, \; S_{n-1}^{(1)}, \; S_{n-1}^{(2)}, \; S_{n-1}^{(3)}, \; S_{n-1}^{(4)}$$

be the tangent spaces (generated by the tangent lines), at O, to the varieties

$$W_{2n}, \; L_n^{(1)}, \; L_n^{(2)}, \; M_n^{(1)}, \; M_n^{(2)} \qquad (*)$$

respectively; they then satisfy the desired conditions for the preceding result, so that they define, as indicated, n projective invariants. Since such spaces undergo homographic transformations only when we subject, in an arbitrary manner, W_{2n} (or V_n and V'_n separately) to a differential (or analytic) transformation, the invariants obtained remain, in fact, unaltered by such transformations. This we indicate by saying that they are *topologico-differential invariants*.

It is now not difficult to recognize that these invariants coincide with those considered towards the end of § 2, p. 4. Those previously introduced by the concept of numbers called coefficients of dilatation of T at a united point, P, are just particular cases, obtainable from the above by assuming V'_n superimposed on V_n and T_2 coincident with the identity transformation of this variety into itself.

We remark, finally, that the preceding results can be applied to two varieties $M_n^{(1)}$ and $M_n^{(2)}$ meeting in a point, O, and immersed in a W_{2n} which has the special property — besides the others which are obvious for the present argument — of *containing two ∞^n systems of varieties $H_n^{(1)}$ and $H_n^{(2)}$ mutually unisecant.* Such systems are in fact, on the $W_{2n} = V_n \times V'_n$ considered above, the systems given by

$$H_n^{(1)} = V_n \times Q', \quad H_n^{(2)} = Q \times V'_n,$$

where Q and Q' are arbitrary variable points on V_n and V'_n respectively. Conversely, if a W_{2n} contains two such systems and, moreover, two varieties $M_n^{(1)}$ and $M_n^{(2)}$ passing in a general manner through a point O, then we can define on $M_n^{(1)}$ a correspondence Θ which transforms a point R of $M_n^{(1)}$ into a point R' of $M_n^{(1)}$ when the $H_n^{(1)}$ through R and the $H_n^{(2)}$

through R' meet in a point on $M_n^{(2)}$. Θ then admits O as a united point; and the n coefficients of dilatation of Θ at O do not differ from the n invariants obtainable in the way just specified starting with (*), when we replace the varieties $L_n^{(1)}$ and $L_n^{(2)}$ by the $H_n^{(1)}$ and the $H_n^{(2)}$ just considered.

§ 4. Local metrical study of the dual transformations. — Let Θ be a transformation of class C^1, which transforms the single points $P(x_1, x_2, \ldots, x_n)$ (of a region) of a real Euclidean E_n into hyperplanes π, with equation $u_1 X_1 + u_2 X_2 + \cdots + u_n X_n + u_{n+1} = 0$, of another Euclidean E'_n $(n \geqq 2)$. Then $u_1, u_2, \ldots, u_{n+1}$ are expressible as functions not all identically zero, with class C^1, of $x_1, x_2, \ldots, x_n$. These functions, manifestly, can still be changed by multiplication with an arbitrary non-zero function of $x_1, x_2, \ldots, x_n$ with class C^1. This would allow us to reduce to unity one non-zero u, and hence to write in a rather simpler — but less symmetric — form some of the analytic developments which follow. We leave the reader to display the different formulations thus obtainable.

Let us now consider the square matrix Δ, of order $n+1$, which is derived from the Jacobian matrix $\dfrac{\partial(u_1, u_2, \ldots, u_n)}{\partial(x_1, x_2, \ldots, x_n)}$ by adding the row

$$\frac{\partial u_{n+1}}{\partial x_1}, \frac{\partial u_{n+1}}{\partial x_2}, \ldots, \frac{\partial u_{n+1}}{\partial x_n}, u_{n+1}$$

and the column

$$u_1, u_2, \ldots, u_n, u_{n+1};$$

and let us put moreover, for brevity,

$$\omega = \sqrt{u_1^2 + u_2^2 + \ldots + u_n^2}\,.$$

We then concentrate on a pair of elements (P, π) — homologous in Θ — for which neither ω nor $\det\Delta$ is zero. This is equivalent to supposing that the hyperplane π is *proper* and that the correspondence Θ is *biregular* in a neighbourhood of this pair.

If a point Q of E_n, near to P, tends to P in an arbitrary manner such that the line PQ tends to a line r, then the hyperplane χ homologous to Q tends to π in such a manner that the space of dimension $n-2$ which is the intersection of π with χ takes up a limiting position, ϱ, depending only on r. The correspondence so obtained, between the lines r of S_n passing through P and the maximal sub-spaces ϱ of π, is a projectivity and is non-degenerate (in view of the biregularity of Θ). This correspondence can then be conveniently completed by establishing, in the obvious way, a biregular correspondence between the orientations of any two homologous elements r and ϱ.

We have, moreover, that the quotient of the infinitesimal angle $\pi\chi$ (measured in radians) and the length of the segment PQ, as $P \to Q$, approaches a (non-negative) limit which depends only on the direction of r; the number and its reciprocal are called, respectively, the *coefficient of dilatation* and the *coefficient of contraction* of Θ at P in the direction r.

With a view to determining the first order differential invariants of the dual correspondence Θ relative to the pair (P, π), we proceed in a way analogous to that followed in § 1, p. 1, for the point transformations. We take on the various semi-lines through P and, starting with this point, a segment equal in length to the associated coefficient of contraction. The variable extremity of such a segment thus generates an elliptical hypercylinder, Γ, which is specialized in a single direction. The axis, a, of this cylinder is called the *directrix* of Θ belonging to the point P; it is precisely the line passing through P which, in the above projectivity between π and the star with centre P, corresponds to the hyperplane at infinity in π. The hyperplane α of E_n perpendicular to a at P meets Γ in a hyperellipsoid, $\mathfrak{J}$, having P as centre. Let A be the (proper) point of π which corresponds to α in the above projectivity.

It is easily verified that such a projectivity transforms any two lines through P conjugate with respect to Γ into two spaces of dimension $n-2$ in π which are perpendicular. Therefore there always exist $n-1$ mutually perpendicular lines $r_1, r_2, \ldots, r_{n-1}$ in α through P which correspond to $n-1$ mutually perpendicular spaces $\varrho_1, \ldots, \varrho_{n-1}$ in π through A of dimension $n-2$. Such lines, called the *principal lines* belonging to P, can also be characterized by means of the property that, corresponding to them, the coefficient of contraction or dilatation has a finite non-zero extremum (not, however, necessarily a maximum or a minimum); to the associated coefficients we also give the qualification of *principal*.

If the above hyperellipsoid $\mathfrak{J}$ has unequal axes, then there is only one $(n-1)$-ple of principal lines through P, given by those axes, which constitute — together with the directrix a — the so-called *principal n-ad* belonging to P. The lengths $1:c_1, 1:c_2, \ldots, 1:c_{n-1}$ of the semi-axes of $\mathfrak{J}$ are manifestly equal to the $n-1$ *principal coefficients of contraction* of Θ at P (so that their reciprocals $c_1, c_2, \ldots, c_{n-1}$ coincide precisely with the principal coefficients of dilatation). It is clear, then, what modifications occur in the foregoing with the hypothesis that $\mathfrak{J}$ is a hyperellipsoid of rotation.

There is also in E'_n a *principal n-ad*, connected with π in an invariant manner, which is rectangular and has A as a vertex; its faces are π and the $n-1$ hyperplanes perpendicular to π and containing separately the spaces $\varrho_1, \varrho_2, \ldots, \varrho_{n-1}$.

Using, for the sake of brevity, the language of infinitesimals, we can say, moreover, that a point Q of E_n infinitely near to P and lying on a is transformed by Θ into a hyperplane χ of E'_n, infinitely near to and parallel to π: the ratio of the distance $\pi\chi$ to the distance PQ is an invariant, c, which we call the *index of elongation* of Θ at P, whilst its reciprocal we call the *index of restriction* of Θ at P. We have that:

A dual transformation Θ between two Euclidean spaces with dimension n, which is biregular and of class C^1, possesses n metrical differential invariants of the first order, given by the index of elongation c and the $n-1$ principal coefficients of dilatation $c_1, c_2, \ldots, c_{n-1}$ of Θ. These n invariants are generally mutually independent, whilst any other metrical differential invariant of the first order is necessarily a function of them.

In order to establish this result, it is convenient to introduce in E_n the unruled hyperboloidal hyperquadric Λ, called the *hyperquadric of deformation* of Θ belonging to P, defined by the following conditions:

1. Λ has P as centre and the n edges of the principal n-ad of Θ belonging to P as its axes;
2. Λ cuts the directrix a (which is its only transverse axis) in two points at a distance $1:c$ from P;
3. Λ cuts the hyperplane α in the hyperellipsoid (of imaginary points) conjugate to $\mathfrak{J}$.

The result enunciated then follows from:

There exists one, and only one, correlation between E_n and E'_n which approximates to Θ in the first order neighbourhood of the pair (P, π), and which transforms the improper hyperplane of E_n into the direction of E'_n perpendicular to π; it also transforms the hyperquadric of deformation Λ belonging to P into the hypersphere of S'_n having centre P and unit radius.

The correlation in question, in order to satisfy this last condition, has, of course, to transform P into π, $r_1, r_2, \ldots, r_{n-1}$ into $\varrho_1, \varrho_2, \ldots, \varrho_{n-1}$ and a into the space at infinity in π. Conversely, these conditions and those previously enunciated, together with that of inducing appropriate orientations on the homologous elements considered, evidently completely characterize such a correlation. Its equations reduce to the simple form:

$$u_1:u_2:\ldots:u_{n-1}:u_n:u_{n+1} = c_1 x_1 : c_2 x_2 : \ldots : c_{n-1} x_{n-1} : 1 : c\, x_n$$

(where the coefficients have precisely the significance previously specified), when E_n and E'_n are referred to the principal n-ads belonging to P and to π, taking the directrix a as the axis x_n and the hyperplane π as the face $X_n = 0$.

§ 5. Calculation of the first order differential invariants just considered. Concerning the analytic determination of the elements previously introduced relative to a transformation Θ, defined in the manner indicated at the beginning of § 4, p. 7, the following results are available. We denote the absolute value of the determinant of the matrix Δ by D, and the algebraic complement of the element $\frac{\partial u_{n+1}}{\partial x_i}$ in this determinant by D_i $(i = 1, 2, \ldots, n)$. Then the direction cosines α_i of the *directrix a* are given by

$$\alpha_i = \frac{D_i}{\sqrt{D_1^2 + D_2^2 + \cdots + D_n^2}} \qquad (i = 1, 2, \ldots, n),$$

and the *index of elongation c* by

$$c = \frac{D}{\omega \sqrt{D_1^2 + D_2^2 + \cdots + D_n^2}}.$$

Let us consider, moreover, the matrix with $n + 2$ rows and $n + 1$ columns which is derived from the Jacobian matrix $\frac{\partial(u_1, u_2, \ldots, u_n)}{\partial(x_1, x_2, \ldots, x_n)}$ by the addition of the rows

$$\frac{\partial \omega}{\partial x_1}, \frac{\partial \omega}{\partial x_2}, \ldots, \frac{\partial \omega}{\partial x_n}, \omega,$$

$$\alpha_1, \alpha_2, \ldots, \alpha_n, 0,$$

and the column formed by the elements

$$u_1, u_2, \ldots, u_n, \omega, 0.$$

We then have that *the sum of the products $t - 1$ at a time of the squares of the $n - 1$ principal coefficients of dilatation* $(2 \leq t \leq n)$, *is obtained by dividing by ω^{2t} the sum of the squares of the $\binom{n}{t-1}\binom{n}{t}$ minors of order $t + 1$ extracted from the above matrix and containing the matrix $\begin{Vmatrix} \omega \\ 0 \end{Vmatrix}$.*

In particular, for $t = n$, we deduce that the *product of the $n - 1$ principal coefficients of dilatation* has the value

$$c_1 c_2 \ldots c_{n-1} = \frac{\sqrt{D_1^2 + D_2^2 + \cdots + D_n^2}}{\omega^n}.$$

Therefore the *product d of the index of elongation c with the $n - 1$ coefficients of dilatation $c_1, c_2, \ldots, c_{n-1}$* is given by

$$d = D/\omega^{n+1}.$$

This differential invariant is called the *density* of the correspondence Θ at the pair (P, π), in so far as — according to the foregoing — it gives a local measure of the compactness near π of the hyperplanes of the space E'_n corresponding in Θ the points of E_n near P.

Another simple geometric significance of the density d is obtained in the following manner. We fix in E'_n any proper point, O, and for any hyperplane π of E'_n (which originates by Θ from a point P of E_n) we consider the foot, P', of the perpendicular from O to it, or else the pole, P'', of it with respect to the hypersphere of centre O and unit radius (so that P' and P'' are inverse points with respect to this hypersphere); assuming (as is allowable) that the point O is the origin of the axes

$(X_1, X_2, \ldots, X_n)$, we obtain the coordinates X_i' and X_i'' of P' and P'' in the form

$$X_i' = -\frac{u_i u_{n+1}}{\omega^2}, \quad X_i'' = \frac{-u_i}{u_{n+1}} \qquad (i = 1, 2, \ldots, n).$$

Recalling also § 1, p. 3, we see from this that:

The density d of the dual correspondence Θ is obtained by dividing by $\overline{OP'}^{n-1}$ the density of the point correspondence which exists between P and P'; or by dividing by $\overline{OP''}^{n+1}$ the density of the point correspondence which exists between P and P''. Hence such quotients are independent of the choice of the point O.

§ 6. Some particular transformations. Relations between densities. — In case we wish to examine more deeply the behaviour of the dual transformation $\Theta: P \to \pi$, between two Euclidean spaces E_n, E_n', it is convenient to introduce the *directrix curves* and the *principal curves*, namely those curves of E_n having as tangent at any point P either the directrix line or one of the $n-1$ principal lines of Θ belonging to P. We can therefore study the special transformations Θ for which one or more of the congruences formed by such curves is e. g. orthogonal, etc.

Particular interest attaches — when $n \geqq 3$ — to those transformations Θ which have, at any point, all the $n-1$ *coefficients of dilatation equal*; this amounts to saying that at any point the hyperquadric of deformation Λ is formed by rotation about the associated directrix. Such dual transformations appear, in certain ways, to be analogous to conformal transformations, in view of the first paragraph of § 2, p. 3, and also taking account of the following characteristic property which we deduce easily from the equations towards the end of § 4, p. 9.

If P and π are a point of E_n and a hyperplane of E_n' which correspond in a transformation of the above type, then to any two directions of E_n, through P and not parallel to the associated directrix, there are associated in π two proper $(n-2)$-dimensional spaces, the angle between which is equal to the angle between the planes projecting the directions under consideration from the directrix.

Without here examining deeply the many questions suggested by the foregoing, we now propose to extend — as far as possible — to the present case some of the investigations previously undertaken for point transformations at the end of § 2, p. 4. To this end, let us consider two dual transformations Θ_1, Θ_2 transforming the points of E_n into the hyperplanes of E_n' and having the pair (P, π) in common. $T = \Theta_1 \Theta_2^{-1}$ is then a point transformation of E_n into itself, which admits P as united point, and which therefore (§ 2, p. 4) is generally endowed with n *coefficients of dilatation* belonging to P. Any one of these invariants is a topologico-differential invariant of T at P, and hence a *differential invariant* of Θ_1 and Θ_2 which belongs to the pair (P, π), and which is not changed by subjecting E_n to a differentiable transformation and E'_n to a homographic transformation chosen arbitrarily. It would be interesting to investigate in which cases such invariants can be expressed as functions of the principal coefficients of dilatation and the index of elongation of Θ_1 and Θ_2 at P (without considering the relative positions of the various principal n-ads). We have a partial result by observing that:

The density of T at P is equal to the ratio of the densities of Θ_1 and Θ_2 belonging to the pair (P, π).

To prove this, we can represent Θ_1 in the way suggested for Θ in § 4, p. 9. It is allowable, moreover, to assume that P is the point $x_1 = x_2 = \cdots = x_n = 0$ and that π is the hyperplane $u_1 = \cdots = u_{n-1} = = u_{n+1} = 0$ (so that for π we have $\omega = \pm u_n$). Further we are allowed to substitute for Θ_2 the associated approximating correlation given by the equations which are deduced from those towards the end of § 4 by putting for $c, c_1, c_2, \ldots, c_{n-1}$ the analogous quantities $c', c'_1, c'_2, \ldots, c'_{n-1}$ belonging to Θ_2. The point correspondence $T = \Theta_1 \Theta_2^{-1}$ has then the equations

$$X_1 = \frac{u_1(x)}{c'_1 u_n(x)}, \ldots, X_{n-1} = \frac{u_{n-1}(x)}{c'_{n-1} u_n(x)}, X_n = \frac{u_{n+1}(x)}{c' u_n(x)}.$$

It suffices, therefore, to calculate the Jacobian determinant of $X_1, X_2, \ldots, X_n$ with respect to $x_1, x_2, \ldots, x_n$ at the point P, and to recall §§ 1, 5, pp. 3,10 in order to establish the result. Thence we deduce immediately that:

The ratio of the densities of two dual transformations at one of their common pairs is a projective invariant.

§ 7. The curvature of hypersurfaces and of Pfaffian forms. — A Pfaffian form

$$\theta = a_1(x)\,d x_1 + a_2(x)\,d x_2 + \cdots + a_n(x)\,d x_n$$

intrinsically defines in the Euclidean E_n of the variables x a dual correspondence Θ; or, more precisely, that correspondence which associates with the point $P(x)$ the hyperplane π with equation

$$a_1(x)(X_1 - x_1) + a_2(x)(X_2 - x_2) + \cdots + a_n(x)(X_n - x_n) = 0.$$

The density of Θ at the pair (P, π) is a metric invariant, ω, which can be called the *curvature* of θ at the point P. In fact, from § 5, p. 10, in the particular case of θ integrable, i. e. of the form

$$\theta = g(x)\, d\{f(x)\},$$

ω reduces to the *total curvature* k at P of the integral hypersurface

$f(x) = $ const. passing through P; this is obtained by a simple calculation, noting that k can be given by the formula below:

$$k = -(f_1^2 + f_2^2 + \cdots + f_n^2)^{-\frac{n+1}{2}} \begin{vmatrix} 0 & f_1 & f_2 & \cdots & f_n \\ f_1 & f_{11} & f_{12} & \cdots & f_{1n} \\ f_2 & f_{21} & f_{22} & \cdots & f_{2n} \\ \cdot & \cdot & \cdot & \cdots & \cdot \\ f_n & f_{n1} & f_{n2} & \cdots & f_{nn} \end{vmatrix},$$

where the lower indices denote differentiation.

Let us consider, secondly, any algebraic hypersurface F of E_n with order $r \geqq 2$. It defines intrinsically a dual correspondence, Θ, transforming any point P of E_n into its polar hyperplane π with respect to F. The density of Θ at the pair (P, π) is then a *metrical invariant* of F at P. When P is a simple point of F, the hyperplane π is that which touches F at P; and it is immediately seen — in virtue of the last but one formula and of § 5, p. 10 — that:

To within the numerical factor $1 - r$, the above invariant then coincides with the total curvature of F at P.

Combining this result with the penultimate theorem of § 6, p. 12, we obtain:

If two algebraic hypersurfaces of E_n touch at a common simple point, P, the ratio of the corresponding curvatures at P is a projective invariant of them. To within a numerical factor (depending only on the orders of the two hypersurfaces), such an invariant is equal to the density at P of the correspondence which associates two points of E_n when they have the same polar hyperplane with respect to the two given hypersurfaces.

The first part of this proposition is extended immediately to the case in which the two hypersurfaces considered are not algebraic. It is sufficient to observe that the ratio of their curvatures at P is equal to that belonging to the two quadrics which osculate them at P; hence the projective invariance of the second ratio follows immediately from that of the first. Such an invariant is called the generalized MEHMKE-SEGRE *invariant.* It arises also — together with the new invariants — in the manner specified by the following theorem, which, on the other hand, is established without difficulty from the last formula and from §§ 5, 6, pp. 10, 12.

Given in E_n two hypersurfaces F, F' which touch at a common simple point P, we fix, in a general manner, a point O and a hyperplane E_{n-1} of E_n, and we consider the correspondence associating two points Q, Q' of F and F' when their tangent hyperplanes form a pencil with E_{n-1}. Such a correspondence transforms P into itself and so induces a correspondence T of E_{n-1} into itself associating $Q_0 = OQ \cap E_{n-1}$ with $Q_0' = OQ_0' \cap E_{n-1}$ and having $P_0 = OP \cap E_{n-1}$ as united point. However, the density of T

at P_0 does not depend on the methods of choosing the point O or the hyperplane E_{n-1}; further it is precisely equal to the ratio of the curvature of F to the curvature of F', both evaluated at P. More generally, each of the $n-1$ topological invariants of T at the united point P_0 (§§ 2, 3, pp. 4, 6) constitutes a projective differential invariant of contact of F and F' at P, depending only on the associated neighbourhoods of the second order.

Historical Notes and Bibliography

The content of §§ 1—6 is taken, essentially, from B. SEGRE [98], with the exception of the last proposition but one of § 6 — which appears here for the first time — and of the last proposition of the same section, which is already to be found (proved rather less simply) in TERRACINI [156]. Also the development of § 3 appears here in full for the first time; but some of the particular cases were already implicit in B. SEGRE [101], and the enunciations of some of the propositions proved here are already presented in B. SEGRE [108]. For a further deeper study of this subject, cf.: B. A. ROSENFELD, Mathem. Sbornik, **24** (66), (1949), pp. 405—428; B. A. ROSENFELD, Trudy seminara po vectornomu i tenzornomu analizu (Moscou), IX (1952), pp. 213—220; A. FUHRMANN, Math. Zeitschr., **62** (1955), pp. 211—240; B. A. ROSENFELD, Les géométries non euclidiennes (Russian), Moscou 1955, pp. 303—304, 675.

The density of a dual correspondence between two euclidean spaces of dimension n has been introduced, by means of an algorithm, by TRICOMI [151, 152], who has used it in various applications. Further, for $n = 2$, it has been set by that author in relation to the probabilistic notions of density relative to a system of points and lines of a plane, for which cf. BLASCHKE [14]; for the case $n = 2$, cf. also TERRACINI [158].

From TERRACINI [156] dates a first geometric interpretation of the above density (less immediate than that which here is yielded by § 5); and to him also we owe the many relations — given in § 7 — between this density and the notion of curvature. For the final proposition of § 7 cf. B. SEGRE [102]. We shall have occasion (in the third Part) to return, from other viewpoints, to the MEHMKE-SEGRE invariant; with reference to this, cf. MEHMKE [79, 80], C. SEGRE [126] BOMPIANI [18, 20], B. SEGRE [92], MASCALCHI [77], and also TERRACINI [158] where applications of the foregoing invariants are made to the theory of linear systems. For some applications of the preceding concepts to the theory of line congruences cf. TERRACINI [157] and B. SEGRE [102], also § 21 and the Notes which follow § 25.

Part Two

Local Properties of Analytic Transformations at their United Points

We shall deal here with certain local properties relative to the *united points of analytic transformations* of a complex variety into itself, and also with the classification to which they give rise. Some of the properties we shall be able to transfer, without difficulty, to the case of the local study of the united points of differentiable transformations of a differentiable variety into itself; but we shall not dwell on this at present.

§ 8. Coefficients of dilatation and residues of transformations in the analytic field. — Let V_n be any complex variety with complex dimension $n \geqq 1$, and let T be a (uniform) analytic transformation of V_n into itself having a united point, O (which is simple for V_n). We shall assume that T is reversible in a neighbourhood of O; this is equivalent to saying that the homography π induced by T among the tangent lines to V_n at O (§ 2, p. 4) is non-degenerate.

We introduce in the neighbourhood of O in V_n any allowable system of internal (complex) coordinates; it will be no restriction to suppose that each coordinate is zero for O. We denote by $(z_1, z_2, \ldots, z_n)$ and $(w_1, w_2, \ldots, w_n)$ the coordinates of two points in such a neighbourhood which correspond in T. Then the equations of T can be written in the form

$$w_h = (c_{h1} z_1 + c_{h2} z_2 + \cdots + c_{hn} z_n) + \cdots \qquad (h = 1, 2, \ldots, n), \quad (1)$$

where the right-hand side is a power series in the z's (with complex coefficients), of which we have written only the terms of the least degree, such that the matrix γ of the c_{ij}'s has determinant $C \neq 0$. Using the present notation, we can write the equations of the tangent homography π in the form

$$\varrho w_h = c_{h1} z_1 + c_{h2} z_2 + \cdots + c_{hn} z_n \qquad (h = 1, 2, \ldots, n; \ \varrho \neq 0), \quad (2)$$

where the z's, w's now appear as homogeneous coordinates of lines in the star of ∞^{n-1} (complex) lines tangent to V at O.

The n coefficients of dilatation of T at O are none other than the latent roots of the matrix γ (§ 2, p. 4), and to them can be attributed the significance of cross-ratios as specified in § 3, p. 6. In view of (2), the ratios of those roots give the *absolute invariants* of π. To such ratios, therefore, can be assigned a simpler and more direct geometric interpretation by means of cross-ratios, in the well known manner which thus makes only the homography π enter the demonstration (cf. BERTINI [13], pp. 80—81).

We shall denote by $\lambda_1, \lambda_2, \ldots, \lambda_n$ the above coefficients of dilatation (or rather the latent roots of γ). They are n complex numbers all different from zero, but not necessarily distinct. We note that for each of these numbers to be different from unity it is necessary and sufficient that O should be a *simple isolated* united point of T. It is in fact immediately seen that, on the product variety $V_n \times V_n$, the variety representing T and that representing the identity transformation have a common tangent line if, and only if, $\lambda = 1$ is one latent root of γ.

When O is a simple isolated united point of T, it is possible to consider the complex number

$$\omega = \sum_1^n {}_h \frac{1}{1-\lambda_h} \, .$$

This is then a topological differential invariant of T at O, which we shall call (for reasons which will appear in the sequel) the *residue of the correspondence T at the point O.*

In order to calculate it, we put

$$\mu = \frac{1}{1-\lambda}, \quad \text{or rather} \quad \lambda = \frac{\mu - 1}{\mu},$$

in the characteristic equation of γ:

$$f(\lambda) = \det [c_{hk} - \lambda \delta_{hk}] = 0, \tag{3}$$

where δ_{hk} has the value 0 or 1 according as $h \neq k$ or $h = k$. Thus we obtain the algebraic equation

$$\mu^n f(\lambda) = \mu^n f\left(\frac{\mu - 1}{\mu}\right) = \det [\mu (c_{hk} - \delta_{hk}) + \delta_{hk}] = 0$$

in μ. The sum of the roots of this equation gives the equality

$$\omega = -\frac{P_1 + P_2 + \cdots + P_n}{P},$$

where P denotes the determinant of order n

$$P = \det [c_{hk} - \delta_{hk}],$$

and $P_1, P_2, \ldots, P_n$ denote its principal minors of order $n - 1$.

§ 9. Transfer to the Riemann variety. — Retaining the notations of § 8, p. 15, we can associate with V_n the relevant Riemann variety (cf. e. g. B. SEGRE [122], § 22); this is a real analytic variety, with real dimension $2n$, and can be extended to the complex field so giving rise to a *complex variety* V^*_{2n}, with (complex) dimension $2n$. The consideration of the latter is equivalent to extending V_n from the complex field to the bicomplex field (for the theory of holomorphic functions of a bicomplex variable cf. SCORZA-DRAGONI [90]).

The image of T on V^*_{2n} is an analytic correspondence, T^*, of V^*_{2n} into itself, which possesses a united point at O^*, the image of O. We shall show that:

The 2n coefficients of dilatation of T^* at O^* are given by the n coefficients of dilatation of T at O and the n numbers which are their complex conjugates.

To this end, separating the real from the imaginary parts, we put

$$c_{hk} = a_{hk} + i b_{hk}, \quad z_h = x_h + i y_h, \quad w_h = u_h + i v_h.$$

Then, from (1), we obtain the equation of T^* in the form

$$\begin{aligned} u_h &= (a_{h1} x_1 + \cdots + a_{hn} x_n) - (b_{h1} y_1 + \cdots + b_{hn} y_n) + \cdots \\ v_h &= (b_{h1} x_1 + \cdots + b_{hn} x_n) + (a_{h1} y_1 + \cdots + a_{hn} y_n) + \cdots \end{aligned} \qquad (h = 1, 2, \ldots, n).$$

Hence, denoting by α and β the matrices of the a_{hk}'s and the b_{hk}'s respectively, and by γ^* the analogue of γ, or rather the Jacobian matrix of the u, v with respect to the x, y evaluated at the point O^*, we obtain for the latter a VOIGT matrix (with real elements):

$$\gamma^* = \begin{Vmatrix} \alpha & -\beta \\ \beta & \alpha \end{Vmatrix}.$$

If we now denote by $f^*(\lambda)$ the determinant of the matrix obtained from γ^* by subtracting λ from the principal elements, then we have the identity

$$f^*(\lambda) = f(\lambda) \cdot \bar{f}(\lambda), \tag{4}$$

where $f(\lambda)$ is the polynomial in λ given by (3) and $\bar{f}(\lambda)$ denotes its complex conjugate. In fact, (4) certainly follows when we put for λ an arbitrary *real* number, by means of a theorem of DRUDE [43] (for which cf. also AGOSTINELLI [1]). This implies that (4) must also hold in the complex field; and from (4) clearly follows the assertion, since the coefficients of dilatation of T^* at O^* are given by the roots of the polynomial $f^*(\lambda)$.

From the theorem just proved follows immediately the trivial fact that the reversibility of T in the neighbourhood of O implies the reversibility of T^* in the neighbourhood of O^*. It follows also that at present O^* is a simple isolated united point of T^*. On this hypothesis, in view of § 8, p. 15, and with the obvious notation, we obtain the expression

$$\omega^* = \Sigma \frac{1}{1-\lambda^*} = \Sigma \frac{1}{1-\lambda} + \Sigma \frac{1}{1-\bar{\lambda}} = \omega + \bar{\omega}$$

for the residue of T^* at O^*. Hence:

The residue of T^* in O^* is always real, and is equal to twice the real part of the residue of T at O.

§ 10. Formal changes of coordinates. — An important problem — to the consideration of which we now pass — is that of investigating how we can simplify equations (1) for T by a suitable choice of coordinates in the neighbourhood of O. The most general admissible change of coordinates, keeping all the coordinates zero at O, is obtained by applying an arbitrary analytic transformation which changes the n-ple $(0, 0, \ldots, 0)$ into itself and which has a non-zero Jacobian for this n-ple. It is clear that in (1) we must operate on z and on w with the same transformation.

Thus, for example, by operating on the coordinates with a suitable linear homogeneous substitution, we can simplify the linear part of equations (1), reducing equations (4) of the tangent homography to the canonical form (cf. for example BERTINI [13], pp. 100—102). We shall limit almost all the subsequent considerations to the case in which the tangent homography is general (and thus possesses n independent united elements). With this hypothesis, which in the sequel — unless the contrary is stated explicitely — we shall always take as understood, and by means of a change of coordinates of the type stated above, we can reduce the equations of T to the following form:

$$\begin{aligned} u &= a\,x + a_{ij\ldots l}\,x^i y^j \ldots z^l \\ v &= b\,y + b_{ij\ldots l}\,x^i y^j \ldots z^l \\ &\ldots\ldots\ldots\ldots\ldots\ldots \\ w &= c\,z + c_{ij\ldots l}\,x^i y^j \ldots z^l . \end{aligned} \tag{5}$$

In (5), $(x, y, \ldots, z)$ and $(u, v, \ldots, w)$ are the new coordinates of two homologous points, and the other letters denote complex numbers satisfying the obvious condition that the power series which appear on the right-hand side — it being understood that the summation extends over the (non-negative integral) indices satisfying

$$i + j + \cdots + l \geqq 2 \tag{6}$$

— converge in an n-dimensional neighbourhood of the point O. In particular, the n coefficients $a, b, \ldots, c$ are none other than the coefficients of dilatation of T at O, so that none of them can be zero.

In general, none of these coefficients can be written in the form:

$$a^i b^j \ldots c^l ,$$

where $i, j, \ldots, l$ again denote n non-negative integers satisfying (6). Indeed, otherwise, there would exist, between the complex numbers $a, b, \ldots, c$, and likewise between their moduli, special relations of an arithmetical nature. We shall then say that at O the correspondence T is *arithmetically special* or *general*, according as there exist or do not exist such relations. In the first case we can consider the *index of speciality*, defined as the number of distinct relation of the above type.

It is clear that (5) are changed into equations of the same kind if, in place of the coordinates $(x, y, \ldots, z)$, we assume those given by the relations

$$\begin{aligned} X &= x + \alpha_{ij\ldots l}\, x^i y^j \ldots z^l \\ Y &= y + \beta_{ij\ldots l}\, x^i y^j \ldots z^l \\ &\ldots\ldots\ldots\ldots\ldots\ldots \\ Z &= z + \gamma_{ij\ldots l}\, x^i y^j \ldots z^l , \end{aligned} \tag{7}$$

which naturally forces us, at the same time, in place of $(u, v, \ldots, w)$ to consider

$$\begin{aligned} U &= u + \alpha_{ij\ldots l}\, u^i v^j \ldots w^l \\ V &= v + \beta_{ij\ldots l}\, u^i v^j \ldots w^l \\ &\ldots\ldots\ldots\ldots\ldots\ldots \\ W &= w + \gamma_{ij\ldots l}\, u^i v^j \ldots w^l , \end{aligned} \tag{8}$$

where the indices $i, j, \ldots, l$ are still non-negative integers satisfying (6), the summation with respect to them being understood, and the various $\alpha, \beta, \ldots, \gamma$ are assigned complex numbers. These numbers will have to satisfy rather exacting restrictions if we want (7) to be considered as representing an effective change of coordinates. This in fact requires that each of the power series appearing on the right-hand side of (7) converges in a n-dimensional neighbourhood of the point O, and will be expressed briefly by saying that the equations (7) *are convergent.* If, however, such conditions are not satisfied, then it will still make sense to consider the right-hand sides of (7) and (8) as formal power series: then from (5), (7) and (8), by eliminating $(x, y, \ldots, z)$ and $(u, v, \ldots, w)$, we shall emerge with unique equations of the type

$$\begin{aligned} U &= a\,X + A_{ij\ldots l}\, X^i Y^j \ldots Z^l \\ V &= b\,Y + B_{ij\ldots l}\, X^i Y^j \ldots Z^l \\ &\ldots\ldots\ldots\ldots\ldots\ldots \\ W &= c\,Z + C_{ij\ldots l}\, X^i Y^j \ldots Z^l\,, \end{aligned} \tag{9}$$

where the right-hand sides will, of course, still be formal power series. We shall, in this case, say that (9) is obtained from (5), near to O, by means of the formal change of coordinates (7).

§ 11. Formal reduction to the canonical form for the arithmetically general transformations. — We propose to prove the following theorem.

If the transformation T is represented in the neighbourhood of O by the equations (5) *and if it is arithmetically general at O, then, given any $A_{ij\ldots l}, B_{ij\ldots l}, \ldots, C_{ij\ldots l}$, there exists one and only one formal change of coordinates of the type* (7) *which gives* (9) *as (formal) equations of T near to O.*

This is equivalent to saying that we must be able to determine in one and only one way $\alpha_{ij\ldots l}, \beta_{ij\ldots l}, \ldots, \gamma_{ij\ldots l}$ such that the following two operations — applied successively — result in identity formulae in $(x, y, \ldots, z)$. Firstly we shall express $(X, Y, \ldots, Z)$, $(U, V, \ldots, W)$ in (9) by means of (7) and (8); this gives immediately

$$\begin{aligned} u + \alpha_{ij\ldots l}\, u^i v^j \ldots w^l &= a\,(x + \alpha_{ij\ldots l}\, x^i y^j \ldots z^l) + \\ &\quad + A_{pq\ldots r}\,(x + \alpha_{ij\ldots l}\, x^i y^j \ldots z^l)^p \times \\ &\quad \times (y + \beta_{ij\ldots l}\, x^i y^j \ldots z^l)^q \ldots (z + \gamma_{ij\ldots l}\, x^i y^j \ldots z^l)^r \\ v + \beta_{ij\ldots l}\, u^i v^j \ldots w^l &= b\,(y + \beta_{ij\ldots l}\, x^i y^j \ldots z^l) + \\ &\quad + B_{pq\ldots r}\,(x + \alpha_{ij\ldots l}\, x^i y^j \ldots z^l)^p \times \\ &\quad \times (y + \beta_{ij\ldots l}\, x^i y^j \ldots z^l)^q \ldots (z + \gamma_{ij\ldots l}\, x^i y^j \ldots z^l)^r \\ &\ldots\ldots\ldots\ldots\ldots\ldots\ldots\ldots\ldots\ldots \end{aligned} \tag{10}$$

$$w + \gamma_{ij\ldots l}\, u^i v^j \ldots w^l = c(z + \gamma_{ij\ldots l}\, x^i y^j \ldots z^l) + \\ + C_{pq\ldots r}(x + \alpha_{ij\ldots l}\, x^i y^j \ldots z^l)^p \times \\ \times (y + \beta_{ij\ldots l}\, x^i y^j \ldots z^l)^q \ldots (z + \gamma_{ij\ldots l}\, x^i y^j \ldots z^l)^r;$$

and, secondly, we shall substitute here for $(u, v, \ldots, w)$ the expressions given in (5). It will then be necessary to satisfy the conditions which could be obtained from the last equations by identifying the coefficients of similar terms on the two sides of each relation.

Now in view of (10) and (5), taking account also of (6), we see immediately that the linear terms furnish no conditions and that — putting $k = i + j + \ldots + l$ ($\geqq 2$) — we obtain from the terms in $x^i y^j \ldots z^l$ conditions of the form

$$\begin{aligned} (a - a^i b^j \ldots c^l)\, \alpha_{ij\ldots l} &= P_{ij\ldots l} \\ (b - a^i b^j \ldots c^l)\, \beta_{ij\ldots l} &= Q_{ij\ldots l} \\ \ldots\ldots\ldots\ldots\ldots & \\ (c - a^i b^j \ldots c^l)\, \gamma_{ij\ldots l} &= R_{ij\ldots l}, \end{aligned} \tag{11}$$

where $P_{ij\ldots l}, Q_{ij\ldots l}, \ldots, R_{ij\ldots l}$ are well defined polynomials in the quantities $a, b, \ldots, c$, in the $a_{pq\ldots r}, b_{pq\ldots r}, \ldots, c_{pq\ldots r}, A_{pq\ldots r}, B_{pq\ldots r}, \ldots, C_{pq\ldots r}$ having weight $p + q + \cdots + r \leqq k$, and, further, (if $k > 2$) also in the $\alpha_{pq\ldots r}, \beta_{pq\ldots r}, \ldots, \gamma_{pq\ldots r}$ with weight less than k. We note, moreover, that the terms of such polynomials, which do not depend on $A, B, \ldots, C$, have for coefficients positive integers and contain $\alpha, \beta, \ldots, \gamma$ to a degree not greater than the first.

Since, by hypothesis, each of the expressions in parentheses in (11) is different from zero, we obtain — by giving to k successively the values 2, 3, 4, ... — from (11), in a recurrent manner and without ambiguity, all the $\alpha, \beta, \ldots, \gamma$ with weights 2, 3, 4, ... Thus the assertion is proved.

From the preceding theorem, by assuming $A, B, \ldots, C$ all zero, we have the following particular case:

Any analytic transformation T, which has, at a united point O, coefficients of dilatation $(a, b, \ldots, c)$ and which is arithmetically general at O, can, by a suitable formal change of coordinates near to O, be formally represented by the linear equations

$$U = aX,\ V = bY, \ldots, W = cZ. \tag{12}$$

However, the equations (12) naturally do not constitute an effective representation of T, if the right-hand sides of equations (7) are not convergent when the coefficients $\alpha, \beta, \ldots, \gamma$ are determined by the procedure just indicated. The convergence of those right-hand sides gives precisely the necessary and sufficient conditions for the effective reducibility of the equations of T to the linear form; with regard to this, it is immediately seen that we do not affect the question if we multiply the coordinates by arbitrary constants all different from zero.

We have of course no conditions of convergence if in the right-hand sides of (7) we consider only the terms for which

$$i + j + \cdots + l \leqq k\,, \tag{13}$$

where k is an integer $\geqq 2$ fixed arbitrarily; in this case, the right-hand sides reduce to polynomials with degree $\leqq k$. The preceding argument, limited to the equations (11) for which (13) holds, shows that:

When T is arithmetically general at O, it is possible, with a suitable algebraic transformation (with degree $\leqq k$, k being any preassigned integer $\geqq 2$), to reduce the equations of T in the neighbourhood of O to the form (9), *where the coefficients $A, B, \ldots, C$, with weight $\leqq k$, have any given values.*

Since k is arbitrarily large, we deduce that:

The only topological-differential invariants of an analytic transformation at one of its united points, at which it is arithmetically general, are those of the first order, that is the coefficients of dilatation.

§ 12. The case of arithmetically special transformations. — The question of the formal reducibility of the equations of T to a given simplified form can also be resolved with the hypothesis that T is arithmetically special at O. For this purpose, it suffices to reproduce — with suitable and essential alterations — the arguments of § 11, p. 20.

In the present case, for certain choices of the non-negative integers $i, j, \ldots, l$ satisfying (6), some of the expressions in parentheses in equations (11) vanish. To make matters concrete, let us e. g. suppose that the expression belonging to the first of these equations is zero. This equation is then simplified by being reduced to the form

$$P_{ij\ldots l} = 0\,; \tag{14}$$

so that — on repeating the procedure of § 11, p. 20 — we have two cases between which to distinguish according as (14) is satisfied or not. Whilst in the first case the same procedure leaves the value of $\alpha_{ij\ldots l}$ undetermined, in the second case the procedure becomes impossible. Thus, *if T has index of speciality equal to one* (§ 10, p. 18), *then the formal reduction of* (5) *to assigned equations* (9) *by means of a formal change of coordinates of the type* (7) *can either be realized in an infinite number of ways or else it is impossible.*

In view of § 11, p. 20, we see moreover that $P_{ij\ldots l}$ contains $A_{ij\ldots l}$ linearly and with coefficient -1. Therefore it is certainly possible to satisfy (14) if the choice of $A_{ij\ldots l}$ is left free. We conclude that:

The formally reduced equations of an analytic correspondence near a united point O, at which the correspondence is arithmetically special with any index, can be taken as (9), *where on the right-hand side $A_{ij\ldots l}$,*

$B_{ij\ldots l}, \ldots, C_{ij\ldots l}$ *are supposed to be zero, except at most for those* $(i, j, \ldots, l)$ *(in number equal to the index of speciality) corresponding to the single relations of the type*

$$a^i b^j \ldots c^l = a, \quad a^i b^j \ldots c^l = b, \quad \ldots, \quad a^i b^j \ldots c^l = c\,.$$

It is evident that one of the non-zero coefficients A, or $B, \ldots$, or C can be reduced to unity by multiplying the new local coordinates on V_n by suitable non-zero constant factors. We can thus choose those factors so that a certain number of the above coefficients are simultaneously reduced to unity. The non-zero coefficients which finally survive then furnish the only topological-differential invariants of T at O which are independent of the coefficients of dilatation.

§ 13. Criteria of convergence for the reduction procedure in the general case. — It was seen in § 11, p. 20, that the effective reducibility of (5) to the linear form (12) is equivalent to the convergence of (7), where the coefficients $\alpha, \beta, \ldots, \gamma$ are obtained from (11) (deduced in the way there indicated using (10)) with the hypothesis that all the $A, B, \ldots, C$ are equal to zero. We shall indicate by (7_0) and (11_0) the equations which are thus deduced from (7) and (11).

In this and the following two sections we shall assign various conditions which are sufficient for the existence of that linear representation, which is tantamount to the above convergence. Moreover, in §§ 15 and 16, pp. 29, 31, we shall give sufficiently general examples for which the convergence does not hold. From all this will clearly follow the essential importance of the arithmetic properties of the coefficients of dilatation for the problem under consideration.

We first prove the following theorem.

An analytic transformation T of V_n into itself, with a united point O and with coefficients of dilatation $(a, b, \ldots, c)$, can — by suitable choice of coordinates in the neighbourhood of O — be represented by linear equations (reducible to the simple form (12)*), if those coefficients satisfy the following conditions:*

I) *each of them has absolute value less than unity;*

II) *T is arithmetically general at O, that is* (§ 10, p. 18) *for any choice of the non-negative integers $i, j, \ldots, l$ satisfying*

$$i + j + \cdots + l = k \geqq 2\,, \tag{15}$$

none of the expressions

$$a - a^i b^j \ldots c^l, \quad b - a^i b^j \ldots c^l, \quad \ldots, \quad c - a^i b^j \ldots c^l \tag{16}$$

vanishes.

In virtue of condition I), by putting

$$\varrho = \max\,(|a|, |b|, \ldots, |c|)$$

we have

$$\varrho < 1\,.$$

Further, if $k \to \infty$, then $a^i b^j \ldots c^l \to 0$. Therefore, because of condition II), and since none of $a, b, \ldots, c$ is zero, the various expressions (16) have absolute value *not less* than a suitably chosen positive real number, σ, which manifestly satisfies

$$\varrho \geqq \sigma .$$

If, for brevity, we put

$$\theta = 1 - \varrho, \quad \tau = \varrho/\sigma , \tag{17}$$

then it follows that

$$\theta > 0, \quad \tau \geqq 1 . \tag{18}$$

We shall simplify somewhat the writing of some of the formulae in the sequel, by denoting respectively by p and q the sums $x + y + \cdots + z$ and $u + v + \cdots + w$. Let us therefore consider the analytic transformation T' (which is also rational) defined by the equations

$$u = \varrho x + H, \quad v = \varrho y + H, \quad \ldots, \quad w = \varrho z + H , \tag{19}$$

where we assume that

$$H = \frac{\varrho \theta p^2}{n(1 - \theta p)} . \tag{20}$$

Expanding H as a power series in $p = x + y + \cdots + z$ and substituting in (19), we obtain for T' the equations

$$\begin{aligned} u &= \varrho x + a'_{ij \ldots l} x^i y^j \ldots z^l \\ v &= \varrho y + b'_{ij \ldots l} x^i y^j \ldots z^l \\ &\cdots\cdots\cdots\cdots \\ w &= \varrho z + c'_{ij \ldots l} x^i y^j \ldots z^l \end{aligned} \tag{5'}$$

(analogous to (5)), where the coefficients $a'_{ij \ldots l}, b'_{ij \ldots l}, \ldots, c'_{ij \ldots l}$ are *positive* real constants, for which it suffices to observe that

$$a'_{ij \ldots l} = b'_{ij \ldots l} = \ldots = c'_{ij \ldots l} \geqq \varrho \theta^{k-1}/n .$$

Taking the equations of T in the form (5), we see from the convergence of these series that there exist positive real constants φ, $\varkappa$ such that, for any choice of the indices $i, j, \ldots, l$ satisfying (15), we have

$$|a_{ij \ldots l}| \leqq \varkappa \varphi^{-k}, \quad |b_{ij \ldots l}| \leqq \varkappa \varphi^{-k}, \quad \ldots, \quad |c_{ij \ldots l}| \leqq \varkappa \varphi^{-k} .$$

Putting for brevity

$$\min\left(1, \frac{\varrho \varphi}{n \varkappa}\right) = \lambda \qquad (\lambda > 0) ,$$

we have, for any integer $k \geqq 2$,

$$\lambda^{k-1} \leqq \frac{\varrho \varphi}{n \varkappa} .$$

It follows that

$$|a_{ij \ldots l} (\lambda \theta \varphi)^{k-1}| \leqq \varkappa \varphi^{-k} (\lambda \theta \varphi)^{k-1} = \varkappa \lambda^{k-1} \theta^{k-1} \varphi^{-1} \leqq \varrho \theta^{k-1}/n \leqq a'_{ij \ldots l} ;$$

and similarly

$$|b_{ij\ldots l}(\lambda\,\theta\,\varphi)^{k-1}| \leqq b'_{ij\ldots l}\,,\ \ldots,\ |c_{ij\ldots l}(\lambda\,\theta\,\varphi)^{k-1}| \leqq c'_{ij\ldots l}\,.$$

If we multiply the coordinates $x, y, \ldots, z$ by the factor $(\lambda\,\theta\varphi)^{-1}$, which of course means that at the same time we must multiply $u, v, \ldots, w$ by the same factor, then we change the series in (5) into analogous series, which appear respectively as contractions of those series which arise in equations (5′) written in the new coordinates. In order that the notation should not be complicated unnecessarily, we may suppose directly that the same *equations* (5) *are contractions of the equations* (5′). Whence, besides

$$|a| \leqq \varrho,\ |b| \leqq \varrho,\ \ldots,\ |c| \leqq \varrho \tag{21}$$

(an immediate consequence of the definition of ϱ), we have

$$|a_{ij\ldots l}| \leqq a'_{ij\ldots l}\,,\ |b_{ij\ldots l}| \leqq b'_{ij\ldots l}\,,\ \ldots,\ |c_{ij\ldots l}| \leqq c'_{ij\ldots l}\,. \tag{22}$$

Now we observe that equations (5′) — that is (19) also — in fact are reduced to the canonical form

$$U = \varrho X,\quad V = \varrho Y,\ \ldots,\quad W = \varrho Z\,,$$

by means of the change of coordinates

$$X = x + \frac{p^2}{n(1-p)}\,,\quad Y = y + \frac{p^2}{n(1-p)}\,,\ \ldots,\ Z = z + \frac{p^2}{n(1-p)}\,, \tag{23}$$

which naturally implies that, at the same time, we have

$$U = u + \frac{q^2}{n(1-q)}\,,\quad V = v + \frac{q^2}{n(1-q)}\,,\ \ldots,\ W = w + \frac{q^2}{n(1-q)}\,. \tag{24}$$

Indeed, from (19), (20) we deduce

$$q = u + v + \cdots + w = \varrho(x + y + \cdots + z) + nH =$$

$$= \varrho p + \frac{\varrho\theta p^2}{1-\theta p} = \frac{\varrho p}{1-\theta p}\ ;$$

so that, from (23), (24), also taking account of (19), (20) and the first of (17), we obtain

$$U = u + \frac{q^2}{n(1-q)} = \varrho x + \frac{\varrho\theta p^2}{n(1-\theta p)} + \frac{\varrho^2 p^2}{n(1-\theta p)(1-\theta p-\varrho p)} =$$

$$= \varrho x + \frac{\varrho\theta p^2}{n(1-\theta p)} + \frac{\varrho^2 p^2}{n(1-\theta p)(1-p)} = \varrho x + \frac{\varrho p^2}{n(1-p)} = \varrho X\,,$$

and similarly

$$V = \varrho Y,\ \ldots,\ W = \varrho Z\,.$$

Expanding the right-hand sides of (23) as power series in $p = x + y + \cdots + z$, we obtain the same transformation in the form

$$\begin{aligned} X &= x + \alpha'_{ij\ldots l}\, x^i y^j \ldots z^l \\ Y &= y + \beta'_{ij\ldots l}\, x^i y^j \ldots z^l \\ &\ldots\ldots\ldots\ldots\ldots \\ Z &= z + \gamma'_{ij\ldots l}\, x^i y^j \ldots z^l \end{aligned} \tag{7'}$$

(analogous to (7)), where the coefficients $\alpha', \beta', \ldots, \gamma'$ are positive real constants, which need not be specified further. We only observe, in this respect, that:

1. equations (7′) are convergent by construction;
2. the above coefficients satisfy the conditions

$$\begin{aligned} (\varrho - \varrho^k)\, \alpha'_{ij\ldots l} = P'_{ij\ldots l}, \quad (\varrho - \varrho^k)\, \beta'_{ij\ldots l} = \\ = Q'_{ji\ldots l}, \quad \ldots, \quad (\varrho - \varrho^k)\, \gamma'_{ij\ldots l} = R'_{ij\ldots l}, \end{aligned} \tag{$11_0'$}$$

analogous to (11_0), with the obvious meanings for the right-hand sides.

We shall show that, however the non-negative integers $i, j, \ldots, l$ satisfying (15) are chosen, we obtain

$$\begin{aligned} |\alpha_{ij\ldots l}| < \tau^{k-1}\, \alpha'_{ij\ldots l}, \quad |\beta_{ij\ldots l}| < \\ < \tau^{k-1}\, \beta'_{ij\ldots l}, \quad \ldots, \quad |\gamma_{ij\ldots l}| < \tau^{k-1}\, \gamma'_{ij\ldots l}. \end{aligned} \tag{25}$$

Thence, from the convergence of (7′), will follow immediately the convergence of (7_0) and so the assertion. We shall now establish (25), proceeding by the method of induction with respect to the weight k of the left-hand sides.

First of all, if $k = 2$, we see immediately from § 11, p. 20 that the first equations of (11_0) and $(11_0')$ are precisely

$$(a - a^i b^j \ldots c^l)\, \alpha_{ij\ldots l} = a_{ij\ldots l}, \quad (\varrho - \varrho^k)\, \alpha'_{ij\ldots l} = a'_{ij\ldots l}.$$

Recalling the definition of σ and equations (22) and (17), we have

$$\begin{aligned} \sigma\, |\alpha_{ij\ldots l}| \leqq |(a - a^i b^j \ldots c^l)\, \alpha_{ij\ldots l}| = \\ = |a_{ij\ldots l}| \leqq a'_{ij\ldots l} = (\varrho - \varrho^k)\, \alpha'_{ij\ldots l} < \varrho\, \alpha'_{ij\ldots l}, \end{aligned}$$

and, therefore, also

$$|\alpha_{ij\ldots l}| < \frac{\varrho}{\sigma}\, \alpha'_{ij\ldots l} = \tau\, \alpha'_{ij\ldots l}.$$

Similarly the other inequalities (25) are established on the present hypothesis.

Let us now suppose that $k > 2$; and let us assume the validity of the relations analogous to (25) for each of the coefficients with weight $2, 3, \ldots, k-1$. Recalling what was said in § 11, p. 20 regarding the polynomials $P, Q, \ldots, R$, and also taking account of (21), (22) and the second equation (18), we have

$$|P_{ij\ldots l}| \leqq \tau^{k-2} P'_{ij\ldots l}, \quad |Q_{ij\ldots l}| \leqq$$
$$\leqq \tau^{k-2} Q'_{ij\ldots l}, \ldots, \quad |R_{ij\ldots l}| \leqq \tau^{k-2} R'_{ij\ldots l}.$$

From the first equations of (11_0) and $(11'_0)$, in view also of the definition of σ, it therefore follows that

$$\sigma\,|\alpha_{ij\ldots l}| \leqq |(a - a^i b^j \ldots c^l)\,\alpha_{ij\ldots l}| =$$
$$= |P_{ij\ldots l}| \leqq \tau^{k-2} P'_{ij\ldots l} = \tau^{k-2}(\varrho - \varrho^k)\,\alpha'_{ij\ldots l} < \tau^{k-2}\varrho\,\alpha'_{ij\ldots l},$$

so that we have finally

$$|\alpha_{ij\ldots l}| < \tau^{k-2}\frac{\varrho}{\sigma}\,\alpha'_{ij\ldots l} = \tau^{k-1}\,\alpha'_{ij\ldots l},$$

which is the first of relations (25). In a similar way we establish the remaining inequalities (25), which completes the proof of the theorem.

It is clear that if, in the neighbourhood of O, T is (reversible and) representable by linear equations, then the same is true of T^{-1}, and conversely. It therefore suffices to apply the theorem just established to T^{-1}, in order to deduce a second similar theorem for which condition II) is unchanged and instead of condition I) put:

I') *each of the coefficients of dilatation of T at O has absolute value greater than unity.*

§ 14. Iteration and permutability of analytic transformations. — Let us now consider an analytic transformation T, for which we retain the notations and hypotheses of § 10, p. 18, such that T can be taken as defined by (5). Applying T successively m ($\geqq 1$) times, we obtain an analytic transformation, T^m, represented by equations of the type

$$\begin{aligned} u &= a^m x + a^{(m)}_{ij\ldots l} x^i y^j \ldots z^l \\ v &= b^m y + b^{(m)}_{ij\ldots l} x^i y^j \ldots z^l \\ &\cdots\cdots\cdots\cdots \\ w &= c^m z + c^{(m)}_{ij\ldots l} x^i y^j \ldots z^l . \end{aligned} \qquad (5^m)$$

This shows that T^m also possesses O as united point, and that *the coefficients of dilatation of T^m at O are the m-th powers of those of T.* Further it is clear that *T^m is representable by linear equations in the coordinates, if this holds for T.*

The converse of this last proposition is not true. Regarding this, it suffices to notice that, for a given $m > 1$ and all values of the indices,

$$a^{(m)}_{ij\ldots l} = 0, \quad b^{(m)}_{ij\ldots l} = 0, \quad \ldots, \quad c^{(m)}_{ij\ldots l} = 0 \qquad (26)$$

does not imply that similarly

$$a_{ij\ldots l}=0,\ b_{ij\ldots l}=0,\ \ldots,\ c_{ij\ldots l}=0\,. \tag{27}$$

Thus, for example, the non-linear transformation

$$u=-c^2x+y^2,\quad v=cy$$

has for its second power the linear transformation

$$u=c^4x\,,\quad v=c^2y\,.$$

We shall show, however, that the converse of the proposition holds under suitable restrictions, expressed in the following theorem.

Let T^m be arithmetically general; in other words, the coefficients of dilatation $(a, b, \ldots, c)$ and the positive integer m are such that, for any choice of the non-negative integers $i, j, \ldots, l$ satisfying (15), *follow*

$$a^m \neq \omega^m,\ b^m \neq \omega^m,\ \ldots,\ c^m \neq \omega^m\,, \tag{28}$$

where, for brevity,

$$\omega = a^i b^j \ldots c^l\,. \tag{29}$$

Then the linearity of T^m implies, in consequence, the linearity of T.

We are going to prove that, in fact, from (26) and (28) follow equations (27). Let us suppose, in order to obtain a contradiction, that all of (26) and (28) hold, but not all of (27). We can consider then one of $a_{ij\ldots l}, b_{ij\ldots l}, \ldots, c_{ij\ldots l}$ which is non-zero, and has the least possible weight; to fix our ideas, let such a coefficient be

$$a_{ij\ldots l}\neq 0\,. \tag{30}$$

Since the coefficient $a_{ij\ldots l}$ has, by hypothesis, a minimum weight for the non-zero coefficients of the non-linear terms appearing in (5), it follows, from the way of obtaining (5^m) and from relation (29), that precisely:

$$a^{(m)}_{ij\ldots l}=a_{ij\ldots l}\,(a^{m-1}+a^{m-2}\,\omega+a^{m-3}\,\omega^2+\cdots+\omega^{m-1})\,.$$

The first equation of (28) shows, moreover, that $a \neq \omega$; thus the equation just obtained can also be written in the form

$$a^{(m)}_{ij\ldots l}=\frac{a_{ij\ldots l}\,(a^m-\omega^m)}{a-\omega}\,,$$

from which follows

$$a^{(m)}_{ij\ldots l}\neq 0$$

in view of (28) and (30). This then contradicts (26) and so proves the assertion.

We note that, if we were to add the hypothesis that $a, b, \ldots, c$ are in modulus all less or all greater than unity, then the preceding theorem would be an immediate consequence of that in § 13, p. 22 together with the final observations in that section. It might be suspected that, also for the validity of the results of § 13, pp. 22, it is possible to weaken the conditions I) and I'); but we shall see (in §§ 15 and 16, pp. 29—34) that this is not so.

From the foregoing we deduce that

A transformation T, which satisfies condition I) or I') and condition II) of § 13, pp. 22, 26, *possesses exactly m^n m-th roots. That is, there exist m^n transformations, $T^{1/m}$, having T as their m-th power; the coefficients of dilatation of those $T^{1/m}$ are given by $(a^{1/m}, b^{1/m}, \ldots, c^{1/m})$ precisely, according to the various possible choices of these n m-th roots.*

We prove this immediately by representing T, as is certainly possible (§ 13, pp. 22, 26), by linear equations similar to (12).

If we managed to establish the existence of a $T^{1/m}$ under conditions less exacting than the above, then we could extend the range of the results of § 13; and it would suffice to ascertain that such a $T^{1/m}$ were representable linearly, in order to be immediately able to conclude that T might be similarly represented.

The problem of transformations which are representable linearly can also be tackled from another viewpoint, at which we shall now glance after having first made some general remarks, themselves not without interest.

From two analytic transformations, T, Θ of V_n into itself, both having O as united point, we derive by repeated applications the products $T\Theta$, ΘT, still operating on V_n and having O as united point. When these two products coincide, we say that T and Θ are *permutable*, a property which, manifestly, is independent of the choice of coordinates. We shall prove that:

If T and Θ are permutable, and if Θ is arithmetically general at O, then from Θ being linearly representable in the neighbourhood of O follows the analogous property for T.

By choosing the coordinates in the neighbourhood of O suitably, we obtain the equations of Θ in the form

$$u = \alpha x, \quad v = \beta y, \quad \ldots, \quad w = \gamma z .$$

Let, for example, the first equation of T, referred to these coordinates, be

$$u = a^{(1)} x + a^{(2)} y + \cdots + a^{(n)} z + a_{ij \ldots l}\, x^i y^j \ldots . z^l,$$

with the usual summation convention. Equating the first coordinates of the transforms of the same point $(x, y, \ldots, z)$ by $T\Theta$ and by ΘT, we obtain the identity

$$a^{(1)} \alpha x + a^{(2)} \beta y + \cdots + a^{(n)} \gamma z + a_{ij \ldots l}\, \alpha^i \beta^j \ldots \gamma^l x^i y^j \ldots . z^l =$$
$$= \alpha (a^{(1)} x + a^{(2)} y + \cdots + a^{(n)} z + a_{ij \ldots l}\, x^i y^j \ldots . z^l) .$$

This, in view of the arithmetic generality of Θ, implies the vanishing of all the $a_{ij \ldots l}$'s, and so the linearity of the first equation of T. In an analogous manner we proceed to the other equations, and thus prove the linearity of T for any system of coordinates in which Θ is representable linearly. Further, by examining the linear terms in the above identity, we deduce that:

If Θ has, at O, coefficients of dilatation which are all different, then the transformations T which are permutable with Θ — in a coordinate system for which Θ assumes a representation having a reduced form — are those, and only those, represented (with the various non-zero coefficients $a, b, \ldots, c$) by

$$u = ax, \quad v = by, \quad \ldots, \quad w = cz .$$

It will be noticed that amongst the above transformations there are, manifestly, an infinite number having at O coefficients of dilatation not satisfying the conditions of § 13, pp. 22, 26. The deduction which we now make from the foregoing therefore aquires some special interest.

Given *any* T with a united point at O, we assume that it is possible to define a Θ possessing the following properties:

1. Θ is permutable with T;
2. Θ is arithmetically general at O;
3. the coefficients of dilatation of Θ at O have their absolute values either all less or all greater than unity.

Then Θ is representable linearly, in view of § 13, pp. 22, 26; so that *the analogous property must hold for* T, from what was said above.

§ 15. On the united points of cyclic analytic transformations. — In the present section, T will denote any analytic transformation of V_n into itself having a united point, O: thus we do not now — as in §§ 8, 10, pp. 15, 17 — make the hypothesis that T should be reversible in the neighbourhood of O, nor shall we suppose that the associated tangent homography at O is general.

In a neighbourhood of O, we can consider the successive powers of T

$$T, T^2, T^3, \ldots;$$

we shall denote respectively by

$$(x_1, y_1, \ldots, z_1), \quad (x_2, y_2, \ldots, z_2), \quad (x_3, y_3, \ldots, z_3), \ldots$$

the transforms, by these transformations, of the generic point $(x, y, \ldots, z)$, the coordinates of which we shall also denote — for uniformity of notation — by $(x_0, y_0, \ldots, z_0)$. It is then clear that T^m transforms the point $(x_h, y_h, \ldots, z_h)$ into the point $(x_{h+m}, y_{h+m}, \ldots, z_{h+m})$.

We say that T is *cyclic of order m*, if — in the neighbourhood of O — T^m is the identity transformation; so that it follows

$$x_h = x_k, \; y_h = y_k, \ldots, z_h = z_k \qquad \text{if } h \equiv k \pmod{m}. \tag{31}$$

If now m denotes the smallest positive integer for which this property holds, then we shall say that T *appertains to the exponent* m.

Since, in any case, at the point O we have

$$\frac{\partial(x_m, y_m, \ldots, z_m)}{\partial(x, y, \ldots, z)} = \left(\frac{\partial(x_1, y_1, \ldots, z_1)}{\partial(x, y, \ldots, z)}\right)^m,$$

and since the left-hand side must reduce to unity in order that T should be cyclic of order m, we see that the right-hand side cannot be zero and so *any cyclic analytic transformation is reversible in the neighbourhood of any of its united points, O.* With the above hypothesis, we can therefore consider the tangent homography, π, of T at O: this then is (reversible and) cyclic, and therefore general (cf. for example BERTINI [13] pp. 79 bis 80 or 104—108). Hence it is now allowable to apply the results of § 14, p. 26, which immediately give:

At a united point of any cyclic transformation of order m, each of the coefficients of dilatation is equal to an m-th root of unity.

We shall enunciate this result more precisely at the end of the present section. Meanwhile, we note that it cannot be inverted; for there manifestly exist transformations T with O as united point, and the coefficients of dilatation all equal to *m-th* roots of unity, without at the same time being cyclic of order m (cf. the beginning of § 16, p. 31). However we choose coordinates in the neighbourhood of O, *such a transformation cannot be represented by linear equations*, since otherwise the transformation T^m would clearly be the identity transformation, contrary to the supposition. We shall end these observations by showing that:

Any cyclic analytic transformation with united point, O, can locally be represented by linear equations for a suitable choice of coordinates in the neighbourhood of O.

In view of the foregoing, if T is cyclic of order m, then we can choose, in the neighbourhood of O, coordinates $(x, y, \ldots, z)$ such that, for $r = 1, 2, \ldots,$ T^r has equations of the form

$$\begin{aligned} x_r &= a^r x + a_{ij\ldots l}^{(r)} x^i y^j \ldots z^l \\ y_r &= b^r y + b_{ij\ldots l}^{(r)} x^i y^j \ldots z^l \\ &\cdots\cdots\cdots\cdots \\ z_r &= c^r z + c_{ij\ldots l}^{(r)} x^i y^j \ldots z^l, \end{aligned} \qquad (32)$$

where $a, b, \ldots, c$ now denote *m-th* roots of unity, whilst the other symbols have their obvious meanings.

Let us now replace $(x, y, \ldots, z)$ by the coordinates $(X, Y, \ldots, Z)$, in the neighbourhood of O, defined by

$$\begin{aligned} X &= (x_0 + a^{-1} x_1 + a^{-2} x_2 + \cdots + a^{-m+1} x_{m-1})/m \\ Y &= (y_0 + b^{-1} y_1 + b^{-2} y_2 + \cdots + b^{-m+1} y_{m-1})/m \\ &\cdots\cdots\cdots\cdots\cdots\cdots\cdots\cdots\cdots \\ Z &= (z_0 + c^{-1} z_1 + c^{-2} z_2 + \cdots + c^{-m+1} z_{m-1})/m . \end{aligned} \tag{33}$$

We note that, from (32), (33) is a transformation of type (7), and that the right-hand sides of (33) converge in an n-dimensional neighbourhood of O. The transform, by T, of $(X, Y, \ldots, Z)$ has new coordinates $(X_1, Y_1, \ldots, Z_1)$ formed from the right-hand sides of (33) by writing $x_1, y_1, \ldots, z_1$ for $x = x_0, y = y_0, \ldots, z = z_0$, and therefore $x_2, y_2, \ldots, z_2$ for $x_1, y_1, \ldots, z_1$, etc. Recalling (31), we obtain the linear equations

$$X_1 = aX,\ Y_1 = bY, \ldots, Z_1 = cZ ,$$

which proves the theorem.

From the result just obtained we conclude immediately that:

A cyclic analytic transformation T of V_n into itself, endowed with a united point O and coefficients of dilatation $a, b, \ldots, c$, appertains to the exponent m if, and only if, m is the least positive integer for which

$$a^m = 1,\ b^m = 1, \ldots, c^m = 1 ;$$

this is equivalent to saying that, when we put

$$a = e^{\frac{2\pi i p}{m}},\ b = e^{\frac{2\pi i q}{m}}, \ldots, c = e^{\frac{2\pi i r}{m}} \qquad \left(i = \sqrt{-1}\right),$$

then the integers $m, p, q, \ldots, r$ must be mutually prime. In order that O should lie in a k-dimensional subvariety of V_n, consisting of united points of T, it is necessary and sufficient that k of the coefficients of dilatation should be unity.

§ 16. Arithmetically general transformations not representable linearly. — It is clear that a T having a united point with coefficients of dilatation all equal to unity, if it is representable linearly, is the identity transformation. Thus:

No analytic transformation, which is not the identity, and which has a united point with coefficients of dilatation all unity, can be represented by linear equations.

This example of a transformation not representable linearly is included amongst those of § 15, p. 30, which all have the relative coefficients of dilatation with unit modulus and which are all arithmetically special (according to the terminology of § 10, p. 18). Still more general examples of arithmetically special transformations, not linearly representable, are easily deduced from § 12, p. 21.

It is of importance to see if there exist arithmetically general transformations which are not linearly representable. To this question we shall reply in the affirmative, by constructing a sufficiently vast class of transformations having the requisite property. For simplicity of exposition, we shall limit ourselves to the case of $n = 2$ variables. But it would not be difficult to extend finally the range of our considerations, and so give examples for any number of variables. We can add that, as a consequence of § 13, pp. 22, 26, for the required transformations the coefficients of dilatation must have absolute values neither all greater nor all less than unity; so that, in particular, for the case $n = 1$ any transformation having a united point at which it is not linearly representable, must have its (unique) coefficient of dilatation in modulus equal to unity.

Let T be an analytic transformation of V_2 into itself, having at the united point O coefficients of dilatation a, b. For simplicity we suppose a, b to be real and positive. Moreover, conforming with an observation made in the previous paragraph, we shall assume the inequalities

$$a > 1, \quad b < 1 .$$

We can then put

$$a = e^{\sigma}, \quad b = e^{-\tau}, \tag{34}$$

where σ and τ are two positive real numbers. We note that T *is arithmetically special or general at O according as the real number*

$$\varrho = \frac{\tau}{\sigma} \tag{35}$$

is rational or irrational.

Indeed, from (34), (35), $a = a^i b^j$ and $b = a^i b^j$ are equivalent, respectively, to $\sigma = i\sigma - j\tau$ and $-\tau = i\sigma - j\tau$, i. e. to $\varrho = \frac{j}{i-1}$ and $\varrho = \frac{j-1}{i}$.

Thus T emerges arithmetically general, if we assume that ϱ is given by an infinite continuous fraction with arbitrary positive integral elements c_i:

$$\varrho = (c_0, c_1, c_2, \ldots) , \tag{36}$$

where the right-hand side converges, as is well known, representing an irrational number ϱ. Moreover, putting $p_{-1} = 1$, $q_{-1} = 0$, denoting $(c_0, c_1, \ldots, c_r)$, written in the form of an irreducible fraction, by p_r/q_r (for $r = 0, 1, 2, \ldots$), and assuming

$$\varrho_r = (c_{r+1}, c_{r+2}, \ldots) ,$$

so that

$$\varrho_r > c_{r+1} ,$$

we obtain

$$\varrho - \frac{p_r}{q_r} = \frac{(-1)^r}{\varrho_r q_r^2 + q_{r-1} q_r} .$$

Hence, for r even, it follows that

$$0 < \varrho - \frac{p_r}{q_r} < \frac{1}{\varrho_r q_r^2} < \frac{1}{c_{r+1} q_r^2} . \tag{37}$$

We assume in (36) that the integers c_i with even indices are arbitrary, whilst for the remainder we impose the single limitation

$$c_{r+1} \geqq q_r^{q_r - 1} \qquad \text{for } r \text{ even}; \tag{38}$$

this manifestly permits a complete definition of ϱ given by (36), by choosing successively (with a certain latitude) the c_{r+1}'s for $r = 0, 2, 4, \ldots$ (i. e. those c which remain to be determined). Now, fixing arbitrarily a positive real number τ, we see that (35) determines σ; so that a, b are defined uniquely by (34). We remark, for future reference, that, in virtue of (37), (38), it follows that

$$0 < \varrho - \frac{p_r}{q_r} < q_r^{-(q_r+1)} \qquad \text{for } r \text{ even}. \tag{39}$$

Let us now consider the analytic transformation, T, defined by the equations

$$u = a x + a_{ij} x^i y^j , \quad v = b y , \tag{40}$$

where the a_{ij}'s $(i + j \geqq 2)$ are real numbers subject to the following conditions (certainly compatible, and not very restrictive):

1. the series $a_{ij} x^i y^j$ converges in a two-dimensional neighbourhood of O;

2. for the particular values of the indices i, j expressible in the form

$$i = q_r + 1 , \quad j = p_r \qquad \text{(with } r \text{ even)}, \tag{41}$$

the a_{ij}'s have a sign of which we defer the choice for a little, and have their absolute values bounded below by a positive constant, which we denote by θ.

From § 11, p. 19, there exists one, and only one, formal change of coordinates of the type

$$X = x + \alpha_{ij} x^i y^j, \quad Y = y ,$$

which makes linear, formally, the equations of T near to O; further, the coefficients α_{ij} which arise here are real numbers defined by the recurrence relations (11), which now reduce to

$$(a - a^i b^j)\, \alpha_{ij} = a_{ij} + \ldots , \tag{42}$$

where the dots stand for real quantities which it is not necessary to specify.

Hence, for the values i, j expressible in the form (41), we shall agree to choose the sign of a_{ij} inductively such that

$$|(a - a^i b^j)\, \alpha_{ij}| \geqq |a_{ij}| ;$$

and this is certainly always possible in view of (42). Therefore, from condition 2. and (41), (34), (35), (39), for the above values of i, j we have successively

$$\begin{aligned} |\alpha_{ij}| &\geqq \theta/\{a\,|1 - a^{i-1}\,b^j|\} = \theta/\{a\,|1 - e^{\sigma q_r - \tau p_r}|\} \\ &= \theta/\{a|e^{q_r \tau(\varrho - p_r/q_r)} - 1|\} > \theta/\left\{a\left(e^{\tau q_r^{-q_r}} - 1\right)\right\}. \end{aligned} \tag{43}$$

We propose to prove that:

The correspondence T defined by (40), where the coefficients a, b, a_{ij} are chosen with the properties just specified, cannot — by any change of coordinates — be represented by linear equations.

In virtue of § 10, p. 19, in order to prove the assertion, it will suffice to show that:

The double series $\alpha_{ij} x^i y^j$, determined in the way indicated above, does not converge in any two-dimensional neighbourhood of the origin.

In order to obtain a contradiction, we suppose that there exist two constants ξ, η, both complex and different from zero, such that the series $\alpha_{ij}\,\xi^i\,\eta^j$ converges. Taking then the real numbers λ, μ satisfying

$$0 < \lambda < |\xi|\,, \quad 0 < \mu < |\eta|\,,$$

we can therefore also choose a positive real constant K such that — for any choice of the indices i, j, and therefore also for those defined by (41) — we have

$$|\alpha_{ij}|\,\lambda^i\,\mu^j < K\,.$$

With such a choice of the indices i, j, and by letting the index r appearing in (41) tend to infinity (so that i, j and q_r also tend to infinity, whilst $\frac{i}{q_r}$ tends to 1, but $\frac{j}{q_r} = \frac{p_r}{q_r}$ and $\frac{j}{i}$ tend to ϱ), we obtain

$$|\alpha_{ij}|^{\frac{1}{i+j}} < K^{\frac{1}{i+j}}\,\lambda^{\frac{-1}{1+\frac{j}{i}}}\,\mu^{\frac{-\frac{j}{i}}{1+\frac{j}{i}}} \to \lambda^{\frac{-1}{1+\varrho}}\,\mu^{\frac{-\varrho}{1+\varrho}} < \infty\,.$$

On the other hand, equations (43) give

$$|\alpha_{ij}|^{\frac{1}{i+j}} > \left(\frac{\theta}{\tau a}\right)^{\frac{1}{i+j}} \left(\frac{\tau q_r^{-q_r}}{e^{\tau q_r^{-q_r}} - 1}\right)^{\frac{1}{i+j}} q_r^{\frac{1}{\frac{i}{q_r} + \frac{j}{q_r}}} \to \infty\,;$$

this contradicts the previous result, and so proves the assertion.

Historical Notes and Bibliography

Almost all the results of the present Part are established here for the first time [cf. also three Notes by B. Segre, *Sui punti fissi delle trasformazioni analitiche*, Rend. Acc. Naz. Lincei, (8) 19 $(1955)_2$, 200—204, 357—361 and (8) 20 $(1956)_1$, 3—7]; yet they have important precedents which we now recall.

The case of $n = 1$, concerning the study of uniform analytic transformations of one variable in the neighbourhood of a united point, may be said to be exhausted by the work on iteration, i. e. the powers of those transformations in a neighbourhood, as above, and on the functional equations (of ABEL and SCHRÖDER) which are connected with this procedure. On this topic, the work of KÖNIGS [58, 59], GRÉVY [53], LEAU [66], LÉMERAY [68, 69] has particularly to be remembered. The results relative to the case of one coefficient of dilatation with unit modulus, have been applied to important questions of stability by LEVI-CIVITA [71] and CIGALA [39], respectively in the cases of the argument of the coefficient commensurable or incommensurable with π.

An accurate study of the transformations of two variables was later made by LATTES [62—65], who did not exclude the possibility of the tangent homography being particular; but he confined himself to the case in which the two coefficients of dilatation were, in modulus, either both greater or both less than unity. And the same restriction is found in the researches of FATOU [46], concerning the functional equations (generalizing those of SCHRÖDER) which appear in such investigations. The case considered in the final proposition of § 16 thus goes well beyond the work of these authors; and it shows what serious difficulties would be presented, should we wish to complete the classification of LATTES [64, 65] for the reduced forms of analytic transformations in two variables in the neighbourhood of a united point. The developments in question have important consequences in the theory of differential equations, which it would be worth while to study more deeply; with reference to this, cf. POINCARÉ [85].

For the argument of § 15, relative to the transformation (33), cf. H. CARTAN [38]; on this question see also BOCHNER [15].

A differential notion which could be usefully employed in the study of an analytic transformation T of V_n into itself, in the neighbourhood of a united point, is the following, which dates from SEVERI [144, n. 130]. A tangent line to V_n at O is called a principal line or principal direction, if it can be obtained as the limit of a line joining two distinct points of V_n, which correspond in T and tend to O. And O is called a *perfect* coincidence, if each tangent line to V_n at O is principal. It can be shown that any isolated coincidence is always perfect (but not conversely), cf. B. SEGRE [116].

The results of § 15 might constitute the fundamentals of a profound study of *birational cyclic transformations*, and of *involutions of an algebraic* V_n; for $n = 2$, cf. the work of GODEAUX [49—52].

Finally we note an interesting question connected with the results obtained towards the end of § 11. In view of those results, any transformation T of S_n into itself, endowed with a united point O at which it is arithmetically general, can be transformed analytically into an analogous T' of S'_n which possesses at O', the homologue of O, an homography to which it approximates to any preassigned order k. The question to be decided, would be *if, and when, the correspondence between* S_n *and* S'_n *which changes* T *into* T' *can be assumed to be birational.* It should be noticed that, in the study of this problem, it is not a restriction to suppose that T itself is birational. It is in fact known (B. SEGRE [115], MANARA [74]), that it is always possible to approximate at O to T, to any order k, with a birational transformation; and it is clearly allowable, in the above problem, to substitute for T such an approximating transformation.

Part Three

Invariants of Contact and of Osculation
The Concept of Cross-ratio in Differential Geometry

In the present Part we shall carry out a systematic study of the *invariants of contact* and *of osculation* relative to curves and varieties, giving for each invariant a simple geometric interpretation. This will have the effect of using in various ways the *concept of cross-ratio*, without leading however to proper extensions of such a notion to the differential field. We shall then consider a number of extensions of this type, methodically relating them to remarkable differential properties of appropriate varieties.

§ 17. Projective invariants of two curves having the same osculating spaces at a point. — Let us consider, in the plane, two branches C, C' of analytic curves having the same centre O and the same tangent t. On the hypothesis that the two branches can be represented simultaneously in the form

$$y = a x^{\mu} + \cdots, \quad y = a' x^{\mu} + \cdots$$

$$(aa' \neq 0, \text{ and } \mu \text{ an integer or fraction} > 1)$$

(for example, if μ is an integer, this condition is equivalent to saying that the two branches have at O the same order of contact, μ, with t; and it is not difficult to specify the geometrical equivalent of this algebraic restriction in general), then the number a/a' is independent of the coordinate system and so constitutes a *projective invariant* of C, C' at O, called a MEHMKE-SEGRE invariant.

The geometric interpretation of these invariants, given at the end of § 7, p. 13, cannot, in general, be applied to the present hypotheses. However, we now have the following even simpler geometric interpretation, due to C. SEGRE [126]. A line r, near to O and not approaching t, meets C, C' and t in the points P, P' and T respectively. If M is any point of r, and if r tends to a limiting position passing through O, then the limit of the cross-ratio $(PP'TM)$ does not depend on the limiting positions of either r or M (which are restricted only by being different from t and O), and is equal to a/a' precisely.

Relying on this result, we can obtain certain *projective invariants of osculation* of two curves in hyperspace, giving at the same time simple *geometric interpretations* for them.

In a projective S_n $(n \geqq 2)$, let L, L' be two branches of analytic curves with the same centre, O, such that the branches have at O the same tangent line, S_1, the same osculating plane, $S_2, \ldots$, and the same osculating hyperplane, S_{n-1}. We assume that L and L' have the same order of contact with each of these osculating spaces. We can then, evidently (and in an infinite number of ways), introduce in S_n projective non-homogeneous coordinates $x_1, x_2, \ldots, x_n$ with origin O so that the branches are represented by equations of the form

$$L)\ x_r = \{a_r x_1^{\mu_r}\}, \quad L')\ x_r = \{a'_r x_1^{\mu_r}\} \quad (a_r a'_r \neq 0;\ r = 2, 3, \ldots, n),$$

where, in the right-hand sides, the brackets indicate that only the terms with the least order of infinitesimals in x_1 are specified, and the μ_r's denote integral or fractional numbers satisfying

$$1 < \mu_2 < \mu_3 < \cdots < \mu_n .$$

We then have the

Theorem. — *The* $n-1$ *quantities*

$$I_r = \frac{a_r}{a'_r} \qquad (r = 2, 3, \ldots, n)$$

are independent of the coordinates, and so remain invariant if L *and* L' *are transformed by any homography of* S_n.

Since the theorem is proved for $n = 2$, in which case — from the second paragraph of the present section — there is also a geometric interpretation for the one invariant, I_2, we shall prove the general result by the method of induction on the dimension n. We suppose therefore that $n \geqq 3$, and that the theorem has already been proved for a space of dimension $n-1$. The required result then follows from the proposition below, which itself is not lacking in interest.

We fix arbitrarily a point P of S_n, which — for a given r — lies in S_{r+1} but not in S_r $(1 \leqq r \leqq n-1)$, and we also fix a hyperplane $\bar{S}_{n-1}$ of S_n which does not pass through P. The two branches $\bar{L}$, $\bar{L}'$, the projections of L, L' from P onto $\bar{S}_{n-1}$, are so situated — at the point which is the projection of O — that it is possible to apply the theorem for the case of a space of dimension $n-1$. We then see without difficulty that:

For fixed r, *the* $n-2$ *invariants* $\bar{I}$ *relative to* $\bar{L}$, $\bar{L}'$ *do not depend on the choice of the point* P *and the hyperplane* $\bar{S}_{n-1}$, *and they are given by*

$$\bar{I}_2 = I_2\,, \ldots, \bar{I}_r = I_r, \bar{I}_{r+1} = I_{r+2}\,, \ldots, \bar{I}_{n-1} = I_n\,.$$

Whence it follows that:

If the two branches L, L' *considered in the above theorem are projected onto any space of* k *dimensions* $(2 \leqq k \leqq n-1)$ *from a vertex* P, *of dimension* $n-k-1$, *which is skew to it and also to the tangent line* S_1, *then two branches are obtained, in the* k*-dimensional space, to which the theorem can be applied. The set of* $k-1$ *invariants of the projections of the two branches is simply obtained, from the set* $(I_2, I_3, \ldots, I_n)$ *of* L *and* L', *by omitting those* I_r *with indices* r *satisfying*

$$dim\,(P \cap S_r) > dim\,(P \cap S_{r-1})\,.$$

By applying this result to the particular case of $k = 2$, we obtain the following property which, from the foregoing, affords an introduction of each of the invariants I_r as a limit of a suitable cross-ratio.

The Mehmke-Segre *invariant relative to the projections of* L, L' *in a plane, from any vertex* P *of dimension* $n-3$ *skew to the plane and to* S_1, *is always a certain one of the invariants* $I_r (r = 2, 3, \ldots, n)$; *and it is precisely that which corresponds to the least* r *for which* $dim\,(P \cap S_r) < r-2$. *This is, of course, the least integer* r *for which* $dim\,(P \cap S_r) = r-3$.

Other geometric interpretations of these invariants are to be found in Vesentini [166].

The relations considered between the branches L, L', and also the corresponding invariants I_r, cannot in general be dualized. However, we can modify them in a simple way, which we shall now specify, when we transform the two branches by some correlation of S_n. In this way we obtain two branches L^*, L'^*, having a common centre O^* (the transform of S_{n-1}), and mutually situated in such a way that the preceding theorem is now applicable. More precisely, for the two transformed branches, we have to *replace the numbers* μ_r *by the quantities*

$$\mu_r^* = \frac{\mu_n - \mu_{n-r}}{\mu_n - \mu_{n-1}} \qquad (r = 2, 3, \ldots, n; \ \ \mu_0 = 0, \mu_1 = 1);$$

and the projective invariants belonging to the transformed branches L^* *and* L'^* *are given by*

$$I_r^* = (I_{n-1}/I_n)^{\mu_r^*} . I_n/I_{n-r} \qquad (r = 2, 3, \ldots, n; \ \ I_0 = I_1 = 1).$$

§ 18. A notable metric case. — Particular interest arises for the case in which, for an Euclidean ambient space S_n of the two branches L, L', the common centre O is an ordinary simple point of each branch. In this case, on the hypotheses of the theorem stated in § 17, p. 37, it follows that (with the previous notation):

$$\mu_r = \mu_r^* = r \qquad (r = 2, 3, \ldots, n).$$

Further we can now consider the $n-1$ curvatures $c_r, c'_r, c_r^*, c_r'^*$ for each of the branches L, L', L^*, L'^* at O. Recalling the foregoing, and applying the generalized FRENET-SERRET formulae, putting also, for brevity,

$$\gamma_r = \frac{c_r}{c'_r}, \quad \gamma_r^* = \frac{c_r^*}{c_r'^*} \qquad (r = 2, 3, \ldots, n),$$

we see that

$$I_r = \gamma_2 \gamma_3 \ldots \gamma_r, \text{ whence } \gamma_r = \frac{I_r}{I_{r-1}} \qquad (r = 2, 3, \ldots, n; \ \ I_1 = 1),$$

and further

$$\gamma_r^* = \frac{I_r^*}{I_{r-1}^*} = (I_{n-r+1}/I_{n-r}) : (I_n/I_{n-1}),$$

or rather

$$\gamma_r^* = \frac{\gamma_{n-r+1}}{\gamma_n} \text{ for } r = 2, 3, \ldots, n-1 \text{ and } \gamma_n^* = \frac{1}{\gamma_n}.$$

Consequently:

If two curves of an Euclidean S_n *have, at a proper point* O *which is an ordinary simple point for them both, the same osculating* r*-space for* $r = 1, 2, \ldots,$ $n-1$, *then the ratio of any corresponding pair of their* $n-1$ *curvatures at* O *is invariant over the homographies of* S_n (which do not transform O into an improper point). *Further these ratios are changed according to the simple rule expressed by the last equations when an arbitrary correlation is applied to* S_n (provided that this correlation does not transform the osculating hyperplane at O into an improper point).

If we pass from the two given curves to their two projections in any proper S_k $(2 \leqq k \leqq n-1)$ *from a general* S_{n-k-1} *(proper or improper), then each of the* $k-1$ *ratios of the corresponding curvatures of the projected curves is equal to the ratio of the curvatures with the same name for the original curves.*

It is immediately seen, from § 17, pp. 37—38, in what way this result should be modified for *special positions* of the vertex of projection.

§ 19. An important extension. — We shall now study a more general situation than that examined in § 18, p. 38 comprising, among others, the case of two curves in ordinary space which touch simply at a point. We shall limit ourselves to the projective aspect, but we could, however, as in § 18, easily deal with the consequent metrical behaviour. We also leave to the reader the problem of seeing whether, in the various cases, the projective invariants here introduced constitute a complete system (in an obvious sense, which can be easily specified).

We fix first two integers n and r which satisfy the conditions

$$n \geqq 3\,, \quad 1 \leqq r \leqq n-2\,.$$

Let us then consider two curves L and L', in a projective S_n, which touch at a point O, which is an ordinary simple point for them both. Let S_k and S'_k be the osculating k-spaces of L and L' at O $(k = 1, 2, \ldots n-1)$: we suppose that

$$S_1 = S'_1,\ S_2 = S'_2, \ldots, S_r = S'_r\,,$$

and that the spaces S_k and S'_k possess no further speciality. In the present case, *the neighbourhoods of order* n *of* O *on* L *and* L' *admit* $n-2$ *independent projective invariants,* which are obtainable as follows.

Firstly, for $r > 1$, we have a set of $r-1$ invariants by projecting L and L' onto their common osculating r-space, S_r, at O from any $(n-r-1)$-dimensional space which is skew to S_r. The projected curves then have $S_1, S_2, \ldots, S_{r-1}$ as their common osculating spaces at O; thus we can apply the theorem of § 17, pp. 37, and obtain their $r-1$ invariants

$$I_2, I_3, \ldots, I_r\,.$$

It is easily verified that these do not depend on the vertex of projection, and so they furnish $r-1$ projective invariants of L, L' at O (the first of which may be called the MEHMKE-SEGRE invariant, since it is an extension of the one so denominated in § 17, p. 36).

In order to define the remaining $n-r-1$ invariants, we make some preliminary observations. By virtue of our hypothesis on S_k and S'_k, denoting by i any of the integers $r+1, r+2, \ldots, n$, we see that the spaces S_{i-1}, S'_{n+r-i} intersect in $S_r = S'_r$ so that their join is a hyperplane, which we denote by ϱ_i. Further we choose in S_n a fixed hyperplane π, not passing through O, and a variable hyperplane χ, also not passing through O, which tends to a limit χ_0 through O, but not containing the line S_1. When χ is sufficiently close to χ_0, the hyperplane χ cuts L and L' in distinct points, Q and Q', uniquely determined in the neighbourhood of O. We denote by P and R_i the points in which π and ϱ_i are cut by the line QQ', and we put

$$\theta_{ij} = (QQ'R_iP)^{n+r-2j+1}/(QQ'R_jP)^{n+r-2i+1},$$

where $i \neq j$ and $i,j = r+1, r+2, \ldots, n$. We can show that, as χ tends to χ_0, θ_{ij} assumes a limit, I_{ij}, which is independent of π and χ_0 and also of the way in which χ approaches χ_0. Each of the $(n-r)(n-r-1)$ numbers I_{ij} so defined is therefore a *projective invariant of contact* of L and L' at O.

Between the members of this set, however, there always exist certain relations. More precisely, we have

$$I_{ij} \cdot I_{ji} = 1$$

and, with the hypothesis that $n \geqq r+3$, if (i, j, h) denote any three different numbers from $r+1, r+2, \ldots, n$, we also have

$$I_{jh}^{n+r-2i+1} I_{hi}^{n+r-2j+1} I_{ij}^{n+r-2h+1} = 1.$$

This shows that precisely $n-r-1$ *of the* I_{ij}*'s are algebraically independent*, such as, for example,

$$I_{r+1,r+2}, \; I_{r+1,r+3}, \; \ldots, \; I_{r+1,n}.$$

Since these are otherwise independent of $I_2, I_3, \ldots, I_r$, we have thus obtained altogether $n-1$ independent projective invariants.

The various results enunciated above can easily be proved analytically, after having introduced suitable coordinates in S_n. This, at the same time, affords simple expressions for those invariants and verifies that they are dependent only on differential elements with order $\leqq n$. We limit ourselves here to indicating that the most convenient non-homogeneous reference system is one in which the origin is O, the axes x_1, $x_2, \ldots, x_r$ are independent lines through O chosen in the spaces S_1, $S_2, \ldots, S_r$ respectively, the axes $x_{r+1}, x_{r+2}, \ldots, x_n$ are independent lines through O, not lying in S_r and situated in the $(r+1)$-dimensional spaces in which $S_{r+1}, S_{r+2}, \ldots, S_n$ respectively cut $S'_n = S_n, S'_{n-1}, \ldots, S'_{r+1}$. With such a system of reference, the equations of L and L' assume the form

$$L)\quad x_2 = \{a_2 x_1^2\}, \ldots, x_r = \{a_r x_1^r\}, x_{r+1} = \{a_{r+1} x_1^{r+1}\}, \ldots, x_n = \{a_n x_1^n\},$$

$$L')\quad x_2 = \{a'_2 x_1^2\}, \ldots, x_r = \{a'_r x_1^r\}, x_{r+1} = \{a'_{r+1} x_1^n\}, \ldots, x_n = \{a'_n x_1^{r+1}\},$$

and we have simply

$$I_k = \frac{a_k}{a'_k} \qquad (k = 2, 3, \ldots, r),$$

$$I_{ij} = \left(\frac{a_i}{a'_i}\right)^{n+r-2j+1} \Big/ \left(\frac{a_j}{a'_j}\right)^{n+r-2i+1} \qquad (i, j = r+1, \ldots, n;\; i \neq j).$$

§ 20. Projective invariants of contact of differential elements of any dimension. — The considerations of §§ 17—19, pp. 36—40 suggest the problem of the *determination of the projective invariants of two or more differential elements of any dimension, mutually situated in any manner in a projective space.* Many particular aspects, full of particular cases, have been dealt with by several authors (particularly BOMPIANI: cf. the Notes following § 25, p. 52); but the final word is yet to be written on the topic. Here, mainly for illustration, we shall confine ourselves to the consideration of some further cases, sufficiently simple, but also quite general.

Firstly we consider, in a projective space S_n with dimension n sufficiently high, two analytic varieties V_m and $\bar{V}_m$, of the same dimension $m \geqq 2$. We suppose that they touch at a point O, which is simple for them both, and at which they have the same r-osculating space, S_a, and the same $(r+1)$-osculating space, S_b $(r \geqq 1,\ m \leqq a < b \leqq n)$, with dimensions a and b such that the difference $b - a$ has the regular value

$$b - a = \binom{r+m}{r+1}.$$

We shall prove that the neighbourhoods of order $r+1$ of the point O on V_m and $\bar{V}_m$ intrinsically determine a *homography* in the star in S_b with vertex S_a; and also that they determine a set of *algebraic cones* with vertex O having dimension $m-1$, order $r+1$, and in number $b-a$ generally, all lying in the space S_m which is tangent to V_m and $\bar{V}_m$ at O. From this we shall easily deduce *canonical representations* of V_m and $\bar{V}_m$ in the neighbourhood of O, and also some *projective* invariants of contact of V_m and $\bar{V}_m$ at O. These last are dependent only on the neigbourhoods of order $r+1$, and can be obtained from the projective invariants of the homography and from the aforesaid cones.

All this we obtain without difficulty by employing the following observations. We consider any S_{b-1} lying in S_b and containing S_a. If $b = n$, such an S_{b-1} is a hyperplane S_{n-1} of S_n; whilst, if $b < n$, it can be cut on S_b (in an infinite number of ways) by an S_{n-1} of S_n — which passes through S_a, but not through S_b — and conversely. In either case, let the S_{n-1} in question cut V_m in a V_{m-1}, which then — from the definitions of S_a and S_b — has at O a point with multiplicity precisely $r+1$: the cone K which is tangent to V_{m-1} at O is an algebraic V_{m-1}^{r+1}, lying in the S_m tangent to V_m at O, which depends only on S_{b-1} (and not on S_{n-1} if $n > b$). As S_{b-1} decribes the ∞^{b-a-1} system of hyperplanes of S_b passing through S_a, the cone K varies in S_m, in a star with vertex O, describing a linear system which is homographic to the above system, and therefore also has dimension $b-a-1$. In virtue of the hypothesis concerning the difference $b-a$, the set of cones K coincides with the linear system of all the V_{m-1}^{r+1}'s of that star. Hence the two linear systems considered are properly defined by the sole knowledge of O, S_m, S_a, and S_b; and the given V_m induces between them a definite (non-degenerate) homography. An analogous homography between them is similarly defined by $\overline{V}_m$. It is now sufficient to multiply the first of these homographies by the inverse of the second, in order to obtain — in the star of centre S_a lying in S_b — a non-degenerate homography, Θ, the $b-a-1$ projective invariants of which furnish similar projective invariants relative to the neighbourhoods of order $r+1$ of O on V_m and $\overline{V}_m$.

With any of the united spaces S_{b-1} of Θ (which, in general, are in number $b-a$) there is associated in S_m, by means of each of the two homographies first considered, the same cone V_{m-1}^{r+1} with vertex O. And the cones so defined are also evidently associated intrinsically with the two neighbourhoods.

We consider, secondly, two analytic varieties V_m, $\overline{V}_m$ of S_n which, for $s = 1, 2, \ldots, k$ (where $k \geqq 2$), have at a point O, simple for them both, the same s-osculating space, S_{d_s}, where the latter has regular dimension

$$d_s = \binom{m+s}{s} - 1 .$$

The preceding results can then be applied $k-1$ times to V_m and $\overline{V}_m$, when r in the above development assumes any of the values $r = 1, 2, \ldots, k-1$. This then provides a certain number of projective invariants and covariants, which depend on the neighbourhoods of O with order $2, 3, \ldots, k$ on V_m and $\overline{V}_m$. We shall prove that:

Each of these neighbourhoods admits a further projective invariant: this supplies altogether $k-1$ projective invariants of contact, mutually independent and independent of those already obtained.

The new invariants can be introduced by the following construction. We take in S_n any one of the spaces S_{n-m-k}, infinite in number, which, for $s = 1, 2, \ldots, k$, cut S_{d_s} in a space S_{d_s-m-s} and not in a space of higher dimension. We project V_m and $\overline{V}_m$ from this S_{n-m-k} onto a general S_{m+k-1}. Thus we obtain two projected varieties, V'_m and $\overline{V}'_m$, which pass simply through the point O', the projection of O, and which have at O' — for $s = 1, 2, \ldots, k$ — the same s-osculating space, with dimension $m+s-1$. We now cut V'_m and $\overline{V}'_m$ with a generic S_k of S_{m+k-1} which passes through O'. Hence we obtain in S_k two curve sections, having an ordinary simple point at O', with the same osculating spaces of dimension $1, 2, \ldots k-1$: these curves, by § 17, p. 37, possess at O' $k-1$ projective invariants, I_r, depending on the neighbourhoods of the r-th order for $r = 2, 3, \ldots, k$.

Unless particular care is taken in choosing S_{n-m-k} and S_k, the number I_r just obtained will depend on those spaces, as well as on V_m

and $\overline{V}_m$. However, it can be verified that I_r is independent of the choice of those spaces if we add the further restriction on the first of them that it should cut S_{d_r} in a S_{d_r-m-r} which lies in a *united* S_{d_r-1} of the homography, Θ_r, which is induced by V_m and $\overline{V}_m$ on the star of S_{d_r} with vertex S_{d_r-1}. This then supplies (for $r=2, 3, \ldots, k$) the *new* $k-1$ *projective invariants* of V_m, $\overline{V}_m$, when we have fixed for each Θ_r *one* of its united spaces S_{d_r-1}. Varying this choice, we shall then add other invariants which are, however, dependent on those already obtained.

We obtain a simple example, in which, with the above notation, we have the values $n=5$, $m=2$, $k=2$, by considering in S_5 *two general surface elements of the second order with the same tangent plane.* From what has gone before, we can then obtain the reduced equations of the surface elements in the form

$$\begin{cases} x_1=\{f_1(x_4, x_5)\} \\ x_2=\{f_2(x_4, x_5)\} \\ x_3=\{f_3(x_4, x_5)\} \end{cases} \quad \text{and} \quad \begin{cases} x_1=\{c_1 f_1(x_4, x_5)\} \\ x_2=\{c_2 f_2(x_4, x_5)\} \\ x_3=\{c_3 f_3(x_4, x_5)\}, \end{cases}$$

where f_1, f_2, f_3 are quadratic forms, and c_1, c_2, c_3 are non-zero constants; whence arise *six projective invariants*, given by c_1, c_2, c_3 and by the cross-ratios of the roots of the quartic equations

$$f_2 f_3=0, \quad f_3 f_1=0, \quad f_1 f_2=0.$$

§ 21. Two applications. — The results of the present Part and of Part One can be extended and correlated by means of suitable geometric considerations; further we can deduce from them new differential invariants, as we shall now show by means of two examples.

We take first any transformation T of S_n ($n \geqq 2$) into itself, such that T has a united point, O. We consider *two different united directions*, r_1 and r_2, of its tangent homography at O (§ 2, p. 4), and denote respectively by λ_1 and λ_2 the coefficients of dilatation of T at O corresponding to those two directions. We can then prove that:

In S_n any curve L through O, which has r as tangent line and $r_1 r_2$ as osculating plane at O, is transformed by T into a curve, L', passing through O and having at that point (not only r, as tangent line, but also) $r_1 r_2$ as osculating plane. Moreover, the Mehmke-Segre *invariant I_2 of L and L' at O* (§ 19, p. 39) *does not depend on the curve L considered, and is equal to $\lambda_1^2 : \lambda_2$ precisely.*

Interchanging in this enunciation the roles of r_1 and r_2, we obtain a geometric interpretation of the ratio $\lambda_2^2 : \lambda_1$, which — together with the foregoing interpretation of $\lambda_1^2 : \lambda_2$ — leads to a way of *expressing the invariants of dilatation of T at O by means of the* Mehmke-Segre *invariants alone.*

Secondly, we consider in a projective S_3 two surface elements F_1 and F_2 of the second order, whose centres we denote by P_1 and P_2 respectively, assuming them to be distinct, and the tangent planes by π_1 and π_2, also assumed to be distinct. We make the further hypothesis that π_1 and π_2 cut in the line, p, which joins P_1 and P_2. This is the well-known situation which arises, in the case of the two focal surfaces of a congruence of lines, in the neighbourhoods of their points of contact with a general line of the congruence. Let Q be any one of the ∞^4 — generally non-degenerate — quadrics containing p, which have π_1 and π_2 as their tangent planes at P_2 and P_1 respectively. We denote by F'_2 the element into which F_2 in transformed by the polarity with respect to Q. It is evident that the centre of F'_2 is P_1 and that its tangent plane is π_1; hence we can consider the MEHMKE-SEGRE invariant of F_1 and F'_2 at P_1. It can be verified that:

This invariant is unaltered by varying the choice of the fixed quadric Q, or by interchanging the roles of F_1 and F_2 in the foregoing; consequently, it is a *projective invariant of the given surface elements F_1 and F_2*, and is *defined symmetrically by them.*

Thus we obtain a simple projective interpretation of a known *projective invariant of a congruence of lines in S_3*.

§ 22. On certain varieties generated by quadrics. — The elementary notion of cross-ratio has repeatedly intervened in a vital manner in several of the preceding differential considerations (cf. for example §§ 3, 17, 19, pp. 5, 36, 40); however, it has always appeared with a certain local character, whilst a proper extension of this concept to the differential field should be applicable to elements not subject to relations of proximity. We shall show in the following sections how an extension of the last type can be made intrinsically, in sufficiently extensive cases, within the orbit of divers differential fields (as, for instance, the projective or the conformal).

With this aim in view, we first consider in a projective S_n — with dimension $n \geqq 3$ — any quadric, Q, and an ∞^1 system (with class C^1) of sub-spaces S_{k+1} $(1 \leqq k \leqq n-2)$ the general member of which does not touch Q, and so meets it in a non-degenerate V_k^2. Thus we obtain an *∞^1 system, Σ, of non-degenerate varieties V_k^2 lying on Q.* We can make correspond to Σ an ∞^k system of curves, said to be *associated with Σ*, characterized by the following two properties:

1. each associated curve lies completely on the variety, W_{k+1}, generated by the ∞^1 varieties V_k^2 of Σ;

2. the tangent spaces to a V_k^2 and one of the associated curves, at any of their common points, are always conjugate with respect to Q.

It is clear that through a general point P of W_{k+1} there passes one, and only one, associated curve, the corresponding tangent at P being uniquely determined by 1. and 2. We propose to prove (by an infinitesimal argument, which could however be rendered completely rigorous: cf. B. SEGRE [103], pp. 19—22), that:

The ∞^k curves associated with Σ always induce a projective correspondence between any two varieties V_k^2 of Σ.

It will suffice to establish this result in the case of two infinitely near V_k^2's. Since, then, it will follow immediately for two V_k^2's at a finite distance — by means of a suitable process of integration — the projectivity between the latter pair being determined as an infinite product of the infinitesimal projectivities which appear successively between two variable consecutive members of Σ, lying between the given pair.

Thus we consider a general V_k^2 of Σ and we denote by $\bar{V}_k^2$ the quadric of Σ infinitely near to it, and by S_{k+1}, $\bar{S}_{k+1}$ the ambient spaces of V_k^2, $\bar{V}_k^2$ respectively. Since, by hypothesis, V_k^2 is non-degenerate, S_{k+1} does not touch Q, and so it possesses a definite polar S_{n-k-2} which is skew to it (and therefore also skew to $\bar{S}_{k+1}$, which is infinitely near to S_{k+1}).

We fix in any manner a point P of V_k^2. The lines which touch Q at P, and are conjugate — with respect to this quadric — to the tangent S_k to V_k^2 at P, are precisely the ∞^{n-k-2} lines which join P to the points of S_{n-k-2}. Amongst such lines there is one, and only one, which cuts $\bar{S}_{k+1}$, given by the intersection of the spaces PS_{n-k-2}, $P\bar{S}_{k+1}$; and thus — by the definition of the associated curves — is the tangent at P to the curve associated with Σ which passes through P. We can therefore say, in infinitesimal language, that the point correspondence set up between V_k^2 and $\bar{V}_k^2$, by means of the ∞^k curves associated with Σ, is subordinate to the perspectivity with vertex S_{n-k-2} between the two ambient spaces S_{k+1}, $\bar{S}_{k+1}$ (skew to S_{n-k-2}): and this is sufficient to prove the theorem.

The homography induced — as was demonstrated — by the curves associated with Σ between any two varieties V_k^2 of Σ, is subordinate to a well defined homography (not in general a perspectivity) between their two ambient spaces. It is evident that, if S_{k+1}, S'_{k+1} and S''_{k+1} are three such spaces, then the homography between S_{k+1} and S''_{k+1} is none other than the product of the homography between S_{k+1} and S'_{k+1} with that between S'_{k+1} and S''_{k+1}. We can therefore consider the trajectories of the homographies between the different pairs of spaces S_{k+1}. We obtain ∞^{k+1} curves ruling the $(k + 2)$-dimensional variety which is generated by the ∞^1 spaces S_{k+1} : these we shall call the *curves associated* with the system Ω formed by those spaces. Amongst these curves are the ∞^k curves associated with Σ; and it is evident that a curve L associated with Ω is also associated with Σ if, and only if, L lies completely on Q, and that this is certainly the case if L has a point in common with Q.

In view of the last paragraph but one, the homography induced by the curves associated with Ω on two consecutive spaces S_{k+1} and $\bar{S}_{k+1}$ of the system is a perspectivity; and it therefore admits the space, S_r, which is the intersection of S_{k+1} and $\bar{S}_{k+1}$ as all consisting of united points. Particular interest arises in the case in which — for any position of S_{k+1} — we have $r = k$, that is when Ω is a developable system of spaces S_{k+1} (or is a limiting case of such a system); this always occurs if $k = n - 2$. With that hypothesis $r = k$, and for any $h = 1, 2, \ldots, k$, two general spaces S_h, $\bar{S}_h$ of S_{k+1}, $\bar{S}_{k+1}$ which are homologues in the perspectivity just considered meet in a space of dimension $h - 1$, given by the intersection (in S_{k+1}) of S_h with $S_r = S_k$. From this we deduce that:

In an S_n, let Q be any quadric, and let Ω be a system of spaces, S_{k+1} $(1 \leq k \leq n-2)$, the general member of which does not touch Q. On the $(k+2)$-dimensional variety generated by Ω, there are then — intrinsically defined — ∞^{k+1} associated curves; and these induce a projective correspondence between any two spaces S_{k+1} of Ω. The general member of this system of curves has no point in common with Q, but ∞^k of them lie entirely on Q. With the hypothesis that Ω consists of the spaces of a developable, $h+1$ general associated curves (where $h = 1, 2, \ldots, k$) determine on the various spaces S_{k+1} of Ω sets of $h+1$ points, joined by spaces S_h which also generate a developable; and the edge of regression of this lies on the variety generated by the $(k-h+1)$-dimensional spaces osculating the edge of regression of Ω.

§ 23. The notion of cross-ratio on certain surfaces. — We shall now obtain a partial answer to the question posed in the first paragraph of § 22, p. 44. This answer will be deduced from the results of that same § 22, for the particular case of $k = 1$. On this hypothesis, a V_k^2 is an irreducible conic, and four points of it determine in the well-known way a cross-ratio. From the analytical view-point, the points of $V_k^2 = V_1^2$ can be put in biregular correspondence with the values of a parameter, defined to within a fractional linear substitution and so called a projective coordinate, such that the cross-ratio of four points of V_1^2 is always equal to the cross-ratio of their corresponding projective coordinates. Recalling the first theorem of § 22, we can thus say that:

If a surface generated by ∞^1 (in general irreducible) conics is immersed in a quadric of S_n $(n \geq 3)$, which does not contain the planes of the conics, then it is always possible to introduce, for the points of the surface, curvilinear coordinates (u, v) such that the curves $u =$ const. and $v =$ const. are the conics and their associated curves respectively, and further such that the parameter v has the significance of a projective coordinate on each conic $u =$ const.

For four points $P_1(u_1, v_1)$, $P_2(u_2, v_2)$, $P_3(u_3, v_3)$ and $P_4(u_4, v_4)$ of a surface as above, such that no three lie on the same associated curve, we can therefore define intrinsically the cross-ratio $(P_1 P_2 P_3 P_4)$ by putting

$$(P_1 P_2 P_3 P_4) = (v_1 v_2 v_3 v_4).$$

This is equivalent to saying that:

$$(P_1 P_2 P_3 P_4) = (P_1' P_2' P_3' P_4'),$$

where P_1', P_2', P_3' and P_4' are the points in which the associated curves through P_1, P_2, P_3 and P_4 meet any of the ∞^1 conics of the surface.

We can then complete the preceding result, by showing that:

On any surface of the above type, the determination of the associated curves can be made dependent on the integration of a Riccati *equation. The conics form part of a doubly conjugate system (or net) if, and only if, any two consecutive conics have two (distinct or coincident) points in common. In this case, the other family of the net is given by the associated curves precisely; and the net has consequently a zero invariant, so that the* Laplace *transform of the net by means of the associated curves degenerates into a curve.*

We now introduce a special case, by assuming $n = 5$. We consider a quadric of S_5 — supposed non-degenerate — which contains a surface of the required type, as a representation (in the manner of Plücker-Klein) of the totality of lines of S_3; then the surface will represent a congruence of lines of S_3 of a special kind. From the known properties of this representation, and from the foregoing remarks, we deduce that:

If a congruence of lines of S_3 consists of ∞^1 reguli, then its ∞^2 lines can be arranged also as ∞^1 ruled surfaces — associated with the reguli — characterized by the property of cutting the reguli harmonically. (It should be recalled that we say that two ruled surfaces of S_3 cut harmonically at a common generator, p, when the projectivity we obtain on p, by making correspond those points at which the two surfaces have the same tangent plane, is an involution. This is, moreover, equivalent to the dual property.) *Such ruled surfaces are obtainable analytically from the integration of a* Riccati *equation. Further they have the property of inducing a projective correspondence between any two reguli of the congruence, which furnishes an intrinsic definition of the cross-ratio of four general lines of the congruence.*

Amongst the congruences of lines of the above type there are, in particular, the congruences W with ruled focal surfaces. The latter are precisely the congruences of lines of S_3 consisting of ∞^1 reguli for which the curves of contact of the reguli with both the focal surfaces of the congruence are asymptotics. These asymptotics are then rectilinear, and the remaining asymptotic curves also correspond on the two focal surfaces, giving the *curves of contact with the ruled surfaces of the congruence which are associated with the* ∞^1 *reguli.*

Any non-developable ruled surface R of S_3 defines, in an intrinsic manner, a line congruence of a type which is a further specialization of that to which we referred in the last paragraph. Such a congruence contains the ruled surface R, and consists precisely of the ∞^1 reguli which osculate R along its various generators. It is in fact known that the two focal surfaces of such a congruence are simply the ruled flecnodal surfaces of R (that is the surfaces generated by the four-point tangents of R).

The ruled surfaces associated with those ∞^1 reguli are ∞^1 ruled surfaces projectively related to R, called the *evolutes* of R; we can obtain them by integrating a RICCATI equation. They allow — in an obvious manner — of the intrinsic definition of the *cross-ratio* of any four generators of R. Further, they supply a way of making the generators of R correspond to a *projective coordinate*, that is to a parameter — defined to within a fractional linear substitution — by means of which that cross-ratio can be expressed in the usual way.

§ 24. Applications to various branches of differential geometry. — The results of §§ 22, 23, pp. 44—48, can be used in many directions in differential geometry, and in the present section we shall give some brief examples.

Let C be a curve of a projective S_3. We suppose that we can associate, in an intrinsic manner, a line p with each point P of C, such that the ruled surface R generated by p is not developable. It then follows that we can define a projective coordinate for the points P of C, given by the analogous coordinate for the generators p of R (§ 23, p. 48). This is equivalent to introducing the notion of a cross-ratio for four points of C. This process can be effected in many ways, corresponding to the different possibilities for the choice of the line p; and the possible relations between the various notions of cross-ratio so obtained supply interesting projective properties of C. We can, for example, take p as the join of P to one of the two HALPHEN points, or to the point of SANNIA, relative to the point P of C (see e. g. FUBINI-ČECH [48], pp. 42—43); etc.

We leave the reader to consider for himself any of the problems suggested by the foregoing. Here we add only one observation, which we shall then be able to use (partially at least) in the following Part, relating to the case in which C is immersed in a given surface, F. We can now take the line p to be any of the lines defined by a neighbourhood of P on F (an asymptotic tangent, the projective normal, an edge of GREEN, etc.). It is clear that the corresponding notions of cross-ratio depend not so much on F as on the projective geometry of C considered as a curve of F.

Other applications follow easily from the final theorem of § 22, p. 46. The first example concerns any ∞^1 system of non-special linear line-complexes in ordinary space. From that theorem we immediately deduce a number of properties relative to such a system by employing (as in § 23, p. 47) the PLÜCKER-KLEIN representation suitably (cf. B. SEGRE [103], n. 25).

The second example, which covers a wide range, and which it would be convenient to extend further, we obtain by assuming the quadric Q and the system Ω of the final theorem of § 22, p. 46 to be, respectively, the absolute quadric of a non-Euclidean S_n and the totality of ∞^1 tangent spaces S_{k+1} to a given V_{k+1} at the points of a generic curve lying on it. Putting $h = k + 1$, we see that:

The geometry of any non-Euclidean space, of dimension $n \geqq 3$, induces a non-Euclidean connection on any proper sub-variety, not isotropic for the space, which has dimension $h \geqq 2$.

The case of curves, which is not included in this theorem, we can deal with by again using the theorem of § 22, and considering the system of quadrics cut by the absolute quadric on the osculating $(k + 1)$-spaces of the curve.

Applications of another type we obtain, finally, by specializing the quadric of S_n — referred to in §§ 22, 23, pp. 44, 46 — to an $(n-1)$-sphere, projecting it stereographically onto a Euclidean E_{n-1}, and then employing the well-known properties of such a projection. Hence we have immediately the following results (in which, for convenience, we have written n in place of $n-1$).

In a Euclidean E_n ($n \geqq 2$), let F be any surface which contains ∞^1 (circular) circumferences. The orthogonal trajectories of these circumferences we obtain by integrating a RICCATI *equation, and they establish between two arbitrary circumferences of F a correspondence which is always projective.*

The orthogonal trajectories of any ∞^1 system of k-spheres of E_n ($2 \leqq k \leqq n-1$) establish between any two spheres a (homographic, or) conformal correspondence.

The correspondences considered in these two theorems possess a further notable property, as a consequence of the last part of the final theorem of § 22, p. 46. However, we shall not here enunciate this property. Instead we observe that the first of these two theorems allows us to define, in the obvious way (cf. § 23, p. 47), the cross-ratio of four points of F. Such a notion is therefore invariant over the MÖBIUS linear fractional transformations in the ARGAND-GAUSS plane if $n = 2$, and over the conformal transformations if $n > 2$. Whence follows the determination of the *cross-ratio of four points* of any curve C of E_n (which is not a minimal curve); this we obtain by applying the theorem to the surface generated by the ∞^1 osculating circles of C. These definitions manifestly have an intrinsic meaning with respect to the circle geometry of the plane and with respect to the conformal geometry of E_n.

§ 25. **Some extensions.** — We shall now show how it is finally possible to extend the results of §§ 22—24, pp. 44—49 in various directions. For this, we first recall some simple properties of projective geometry.

Two conics lying in different planes and touching at a point are in perspective in one, and only one, way. The perspectivity in question transforms their point of contact into itself, and associates two distinct points if, and only if, the tangents at those points meet on the common tangent of the two conics; this common tangent consists of the united points of the perspectivity.

In an S_n of dimension $n \geqq 3$, we now consider ∞^1 conics which osculate a non-planar curve at its various points. Two consecutive curves of this set are then mutually situated in a way similar to that described in the previous paragraph. Consequently, by using the properties recorded above, and arguments analogous to those employed in §§ 22, 23, pp. 45, 47, we see that:

∞^1 conics osculating a non-planar curve, C, intrinsically define ∞^1 curves associated with them and characterized by the property of forming a net with them. The tangents to the associated curves at the points of a conic always generate a quadric cone, such that the LAPLACE *transform of the net with respect to the associated curves is degenerate. A point-point correspondence is induced on any two of the associated curves by the conics, in such a way that the lines joining homologous points generate a ruled developable surface; and the edge of regression of this surface lies on the ruled surface generated by the tangents of C. Moreover, a projective correspondence is induced between any two conics by the associated curves. Whence a projectively invariant definition is obtained for the cross-ratio of four general points of the surface generated by the conics, and so also, in particular, of four points of the curve C.*

From the above theorem, in order to define the *cross-ratio of four points of a curve C in a projective $S_n (n \geqq 3)$*, it is sufficient to associate — in a projectively invariant manner — with the general point P of C a conic which osculates C at P. This we arrive at in the simplest way as follows. If n ($\geqq 3$) is the dimension of the ambient space of C, then there exists one, and only one, rational normal curve, Γ_P, of S_n which passes through P and the $n+2$ points following P on C. The ∞^1 hyperplanes which osculate Γ_P meet the osculating plane, π, of C at P in lines which envelop a conic with the required properties (this conic is also obtainable as locus of the traces on π of the spaces S_{n-2} which osculate Γ_P at its various points).

The results so arrived at can be easily extended further, if we use — instead of the elementary properties noted in the second paragraph of this section — the following generalized properties.

We fix, in any manner, two integers, k and n, which satisfy $1 \leqq k \leqq \leqq n-2$, and we consider in S_n two quadrics of dimension k, V_k^2 and $V_k'^2$, both non-degenerate, such that their ambient spaces, S_{k+1} and S'_{k+1}, intersect in an S_k which meets V_k^2 and $V_k'^2$ in the same quadric U_{k-1}^2. It is clear that U_{k-1}^2 cannot be more than simply degenerate. If U_{k-1}^2 is degenerate, then there exists one, and only one, perspectivity between S_{k+1}, S'_{k+1} which transforms V_k^2 into $V_k'^2$. If, on the other hand, U_{k-1}^2 is not degenerate, then there exist exactly two such perspectivities; the vertices of these perspectivities are then two points of the S_{k+2} joining the spaces S_{k+1} and S'_{k+1}, which form a harmonic set with S_{k+1} and S'_{k+1}. Consequently, if V'_k approaches V_k, so that S'_{k+1} approaches S_{k+1}, one of the vertices will tend to a position in S_{k+1}, whilst the limiting position of the other will not in general lie in S_{k+1}. Using infinitesimal language, we can therefore say that, when V_k^2 and $V_k'^2$ are infinitely near, then one, and only one, of the two perspectivities of the general case is non-degenerate.

Utilizing the preceding remarks, we can — with arguments similar to those used in § 22, p. 45 — prove the following theorem.

Let W_{k+1} be any variety of S_n $(1 \leqq k \leqq n-2)$ which consists of ∞^1 varieties V_k^2, generally non-degenerate, whose ambient spaces are the osculating spaces S_{k+1} of a curve C (or, more generally, spaces constituting an ∞^1 system of a developable character). Further suppose that the intersection of W_{k+1} with each osculating S_k of C is precisely the V_{k-1}^2 in which S_k meets the V_k^2 lying in that osculating S_{k+1} of C which contains S_k. Then on W_{k+1} there is intrinsically defined an ∞^k system of associated curves, which are characterized by the property that the tangents to these curves at the points of any single V_k^2 generate a (k-dimensional) quadric cone. Moreover, the ∞^1 quadrics V_k^2 are set two by two in projective point-point correspondence by the ∞^k associated curves.

The system of associated curves can be amplified (in one and only one way) to an ∞^{k+1} system, lying on the variety generated by the spaces S_{k+1}, and inducing between any two such spaces the homographic correspondence which subordinates between their V_k^2's the point-point correspondence defined above. Consider $h+1$ general members of this system of curves (such curves could well be also associated curves), where h has one of the values $1, 2, \ldots, k$; the sets of $h+1$ points determined by them in the various spaces S_k are joined by spaces S_h generating a developable, the edge of regression of which lies in the variety generated by the $(k-h+1)$-osculating spaces of C.

Employing the preceding results, we could deal with many allied problems; but here we shall not attempt to do sò. Some of the above results it should be possible also to extend — modified accordingly — to varieties (in general not algebraic) *which are generated by infinite systems of algebraic varieties satisfying suitable conditions* of a type more comprehensive than those previously considered.

Historical Notes and Bibliography

The invariants I_r, of § 17, were introduced in B. SEGRE [96], where are derived the various properties which are exhibited in §§ 17, 18, pp. 36—39; see also VESENTINI [166]. A very special case of those invariants was previously considered in C. SEGRE [130]. On this topic see also SU [152], where there is given a definition of the invariants $I_2, I_3, \ldots, I_r$ rather less simple than that at the beginning of § 19, p. 39. The invariants I_{ij}, which also arise in § 19, p. 40, originate from B. SEGRE [109]; for the particular case of $r = 1$, they were also considered in HSIUNG [54].

Further properties of pairs of curvilinear differential elements have been obtained by A. TERRACINI in the following recent papers: Rendic. Semin. Mat. Torino, 12 (1952/53), 265—281; Bollettino Un. Mat. Ital., (3) **8** (1953), 368—374; Atti Acc. Sc. Torino, **88** (1953/54), 7—15; Rendic. di Mat. e delle sue Appl., (5) **14** (1955), 439—454.

The results of § 20, pp. 41—43 are taken from B. SEGRE [112]; previously SU [153] obtained, by an algorithmic process, the $k-1$ projective invariants of contact, introduced in the penultimate paragraph of § 20, without, however, assigning any geometric interpretation to them.

For the two examples of § 21, pp. 43—44, cf. B. SEGRE [102]. The invariants of a congruence of lines of S_3, given with a quite simple projective meaning at the end of § 21, were first introduced with a metrical procedure by WAELSCH [167, 168]; cf. also BUZANO [35]. Rather complicated projective interpretations are to be found in BOMPIANI [21] and TERRACINI [157].

The study of the various cases which can arise from two surface elements of the second order in ordinary space, together with the discovery of a complete system of projective invariants and their metrical significance, is to be found in BUZANO [36]; cf. also BOMPIANI [28, 29, 31, 32], to whom we are indebted for further extensions in many directions. One of the outstanding merits of the last author is that of having introduced infinitesimal projective invariants of differential elements; this concept, with others, assisted him in a suggestive reconstruction of the projective differential geometry of a surface as first obtained by FUBINI: cf. the *Appendix II* of FUBINI-ČECH [48], and the works there cited, as well as what appears later in § 29, p. 60.

In §§ 22—24, pp. 44—50 we reproduce — with some omissions and variations — the results contained in B. SEGRE [103]. However, it should be added that the introduction of the evolutes of a non-developable ruled surface, and the consequent definition of the cross-ratio of four generators of such a surface (§ 23, p. 48), were previously given in E. CARTAN [37] by a direct algorithmic method. Moreover, the last proposition but one of § 24, p. 49, in the special case of a surface in ordinary space, is to be found already in DARBOUX [40], p. 145, and B. SEGRE [91]; the second of these works also gives the final proposition of § 24 in the case of ordinary space, together with the further properties at which we hinted in the paragraph following the enunciation of that proposition.

It should be noted that, when we speak of the "cross-ratio" of four points of a curve, we must not merely think of some "invariant" of the set of four points and the curve (as, for example, does MARLETTA [76]; cf. in this respect BOMPIANI [30]). Indeed, the invariant for which we are looking must satisfy the following condition: if we take any five points of the curve, then the invariants of the various sets of four of them have to verify those relations which are obtained, in an obvious way, from the ordinary notion of cross-ratio relative to five points of a line. The generalized definition of a cross-ratio, understood in this way, is equivalent to the introduction, on the curve, of a "projective coordinate" (with the significance given to this expression in § 23, p. 46); that is, the introduction of a parameter which has an intrinsic significance and which is defined by this to within a linear fractional substitution precisely. From this point of view, the problem is intimately connected with that of the normalization of ordinary differential equations, which has been dealt with by LAGUERRE and FORSYTH in their well-known works. The connection between the two problems has been thoroughly examined in WILCZYNSKI [171] and, particularly, in BOMPIANI [23, 24], where are to be found — established by analytical methods — the properties considered in § 25, p. 50 as consequences of the general theorem obtained first in that section.

All the other results of § 25 appear here for the first time.

Part Four

Principal and Projective Curves of a Surface, and Some Applications

By applying several of the results of the preceding Part, we shall now introduce — on any non-developable surface of an ordinary space — some important *systems of curves*, the study of which will enable us to extend in certain directions the projective-differential theory of surfaces.

§ 26. Some results of projective-differential geometry. — We begin by recalling some classical results of projective-differential geometry (for which cf. FUBINI-ČECH [48], § 16 A), D) and pp. 99, 141, also § 29, p. 59 which follows in the present work).

We can define a surface F of a projective S_3 by expressing the homogeneous coordinates of a variable point, $x = (x_1, x_2, x_3, x_4)$, of it as functions of two parameters u, v, which thus act as internal curvilinear coordinates on F. We shall suppose that F is non-developable, so that the asymptotics of F may be assumed to be the curves u, $v =$ const. (and this implies that, if we wish to remain in the real field, we must limit ourselves to a portion of F consisting of hyperbolic points). Then the x_i's satisfy a (completely integrable) system of differential equations of the type

$$x_{uu} = \theta_u x_u + \beta x_v + p_{11} x, \qquad x_{vv} = \gamma x_u + \theta_v x_v + p_{22} x, \tag{1}$$

where the suffixes u and v signify differentiations with respect to these variables. Conversely, every such system of equations determines, to within a homography, a surface F referred to its asymptotics.

If we denote by $\pm a^2$ the determinant

$$\pm a^2 = (x\, x_u\, x_v\, x_{uv}) \tag{2}$$

(where the sign on the left-hand side is the sign of the determinant itself), then the function θ appearing in (1) satisfies the equation

$$e^\theta = |a| \, . \tag{3}$$

If we now put

$$F_2 = 2\,a\,du\,dv\,, \quad F_3 = a\,(\beta\,du^3 + \gamma\,dv^3)\,,$$

then the ratio F_3/F_2 is the *linear projective element* of F (according to FUBINI). The curves of F along which the first variation of the integral of this ratio is zero, are the integrals of the differential equation

$$2\,(\beta\,w^3 - \gamma)\,w' = \beta_u w^5 + 2\,\beta_v w^4 - 2\gamma_u w^2 - \gamma_v w\,, \tag{4}$$

where, for brevity, we have put

$$w = \frac{du}{dv}\,, \quad w' = \frac{dw}{dv} = \frac{d^2u}{dv^2}\,, \tag{5}$$

and they are the so-called *pangeodesics* of F (cf. FUBINI-ČECH [48], p. 99, formula $(20)_{\text{ter}}$ and p. 141, where, however, the formula (4) is given with a troublesome error in sign). They are indeterminate if, and only if, ($\beta = \gamma = 0$ or rather) F is a quadric; otherwise they constitute an ∞^2 system, which is covariant over the homographies and over the correlations of S_3.

If F is not ruled (that is, if $\beta\gamma \neq 0$), then the coordinates x_i can be normalized (i. e. changed by the same factor, which is taken as a suitable function of u, v) such that $a = \beta\gamma$; then F_2 and F_3 are reduced to the *normal forms of* FUBINI, which we denote — as usual — by φ_2 and φ_3 respectively.

Whatever choice is made for the factor of proportionality of the x_i's, the tangent plane of F can be defined by the PLÜCKER coordinates $\xi = (\xi_1, \xi_2, \xi_3, \xi_4)$, such that

$$(\xi\,\xi_u\,\xi_v\,\xi_{uv}) = (x\,x_u\,x_v\,x_{uv})\,. \tag{6}$$

Hence the ξ_i's are the integrals of a system of equations analogous to (1), and these equations are precisely

$$\xi_{uu} = \theta_u\,\xi_u - \beta\,\xi_v + \pi_{11}\,\xi\,, \quad \xi_{vv} = -\gamma\,\xi_u + \theta_v\,\xi_v + \pi_{22}\,\xi\,, \tag{7}$$

where

$$\pi_{11} = p_{11} + \beta_v + \beta\theta_v\,, \quad \pi_{22} = p_{22} + \gamma_u + \gamma\,\theta_u\,. \tag{8}$$

It is convenient, moreover, to introduce the self-dual quantities

$$q_{11} = p_{11} + \pi_{11} = 2p_{11} + \beta_v + \beta\theta_v\,, \quad q_{22} = p_{22} + \pi_{22} = 2p_{22} + \gamma_u + \gamma\theta_u\,. \tag{9}$$

By interchanging, if necessary, the names of the u, v curves, and by normalizing the coordinates suitably, we can always obtain the case in which

$$(x\,x_u\,x_v\,x_{uv}) = \frac{1}{2}\,; \tag{10}$$

then (2), (3) and (9) give

$$a = \frac{1}{2}\sqrt{2}\,, \quad \theta = -\frac{1}{2}\log 2\,, \quad q_{11} = 2p_{11} + \beta_v\,, \quad q_{22} = 2p_{22} + \gamma_u\,. \tag{11}$$

Therefore, with the hypothesis that (10) holds, the conditions of integrability for (1) reduce to

$$\begin{gathered} q_{11v} = 2\beta\gamma_u + \gamma\beta_u, \quad q_{22u} = 2\gamma\beta_v + \beta\gamma_v, \\ \beta q_{22v} + 2q_{22}\beta_v - \beta_{vvv} = \gamma q_{11u} + 2q_{11}\gamma_u - \gamma_{uuu}. \end{gathered} \tag{12}$$

However the x_i's are normalized, provided the ξ_i's are chosen to satisfy (6), the differential equation (4) of the pangeodesics can be put in the form

$$(x\ dx\ d^2x\ d^3x) = (\xi\ d\xi\ d^2\xi\ d^3\xi), \tag{13}$$

which permits the writing of such an equation in any curvilinear coordinates (u, v) (also for those coordinates which are not asymptotic). It should be noted that, although (13) (as also (4)) is in fact a second order differential equation, each of the equations

$$(x\ dx\ d^2x\ d^3x) = 0, \quad (\xi\ d\xi\ d^2\xi\ d^3\xi) = 0 \tag{14}$$

is of the third order. The curves which are the integrals of (14) are, respectively, the plane sections and (dually) the so-called cone curves of F.

§ 27. The definition and main properties of the principal and projective curves. — Now let L be any (non-singular) curve of a non-developable surface, F, of S_3; and let O be any point of L which is simple for both L and F. There exists precisely one of the ∞^3 cone curves of F which passes through O and the two points which follow O on L. Calling such a curve C, we see that evidently O is, in general, an ordinary simple point of each of C and L, and further that C and L have — by construction of C — the same tangent line and osculating plane at O. As a consequence of the first theorem of § 17, p. 37, C and L — taken in that order — admit at O two projective invariants I_2 and I_3, with a known geometrical significance. A simple calculation will show that the first of these has always the numerical value 1, whilst the second — which we shall henceforward denote simply by I — in general depends effectively on the third order neighbourhoods of the point O on L and F. More precisely, relying on the developments of § 26, p. 54 (keeping the notation used in that section), and supposing, as is allowable, that (6) holds, we deduce that

$$I = \frac{(x\ dx\ d^2x\ d^3x) - (\xi\ d\xi\ d^2\xi\ d^3\xi)}{(x\ dx\ d^2x\ d^3x)}, \tag{15}$$

where the derivatives are taken along L, and the right-hand side is, of course, evaluated at O. From the definition of I, or else from (15) and § 26, p. 55, we see that $I = 0$ *for every curve of* F *if, and only if,* F *is a quadric*; we shall suppose, therefore, from now on in the present section, unless the contrary is explicitly stated, that F is not a quadric.

The form of equation (15) itself proves immediately that I *is a projective invariant*; this also clearly follows from the way that I was first introduced, which gives a simple projective interpretation for I (and also a metrical interpretation, if this is desired, by means of § 18, p. 38).

An analogous *projective invariant of immersion of L in F* can readily be obtained if we dualize the foregoing, that is by considering instead of C the plane section of F which osculates L at O, etc. Denoting the new invariant by I^*, as a consequence of (15) we have

$$I^* = \frac{(\xi\, d\xi\, d^2\xi\, d^3\xi) - (x\, dx\, d^2x\, d^3x)}{(\xi\, d\xi\, d^2\xi\, d^3\xi)}, \tag{16}$$

and we can also obtain a simple geometric meaning for this invariant.

Connecting the invariants I and I there is the following important relation*

$$I I^* - I - I^* = 0, \tag{17}$$

which is an immediate consequence of (15) and (16). Further, taking account of the geometric meaning of (13) and (14), we see that the plane sections, the cone curves, and the pangeodesics of F can be characterized by the conditions

$$I = \infty \ (\text{or } I^* = 1); \quad I = 1 \ (\text{or } I^* = \infty); \quad I = 0 \ (\text{or } I^* = 0)$$

respectively. The first and second relations are evident, *a priori*, from the definitions of I^* and I respectively. The third result is equivalent to the following:

A curve L of a surface F is a pangeodesic if, and only if, for each point O of L the cone curve of F which osculates L at that point possesses a stationary osculating plane at O; and dually.

More generally, from (15)-(17), we have immediately that:

If one of the invariants I, I remains constant along a curve L of F, then so also does the other. Conversely, if we assign to I and I* non-zero values satisfying* (17), *then there exist on F* ∞^3 *curves along which I and I* assume these values; and they are given by the curves,* L_k, *which are integrals of the differential equation*

$$(x\, dx\, d^2x\, d^3x) = k\, (\xi\, d\xi\, d^2\xi\, d^3\xi), \tag{18}$$

where k is (a constant $\neq 1$) *such that* $I = 1 - 1/k$, $I^* = 1 - k$.

Such a curve L_k will be called briefly a *principal curve, of index k,* of F. For every $k \neq 1$ the curves L_k form an ∞^3 system, which is clearly *covariant over all the homographies*. In particular, the curves L_0 are the plane sections and the curves L_∞ are the cone curves of F. The curves L_1 — the integrals of the equations (18) when $k = 1$ — form only an ∞^2 system, and coincide with the pangeodesics. From (18), if we apply any correlation to F, then every L_k is transformed into a $L_{1/k}$. Therefore if, and only if, $k = 1/k$ the curves L_k are *covariant over the homographies and the correlations*; hence this occurs for $k = 1$, that is — as we have already seen (§ 26, p. 54) — for the pangeodesics, and for $k = -1$, which furnishes an ∞^3 system of curves of F having the above property of projective covariance. We shall call the curves L_{-1} of this system the *projective curves of F*. It is evident, from the foregoing, that:

The projective curves of a surface of S_3 (which is neither a ruled developable nor a quadric) are characterized by the property of having the invariant $I = 2$, or $I^* = 2$, at every point; and this furnishes two different geometric definitions for them (which are mutually dual).

It follows that:

The totality of the principal curves of a surface of S_3 (which is neither a ruled developable nor a quadric) is a continuous ∞^4 system, which is invariant over the homographies and over the correlations.

On the other hand, for a quadric, the plane curves, the cone curves and the projective curves reduce, manifestly, to the same ∞^3 system.

§ 28. Further properties of the above curves. — The properties just discussed (in § 27, p. 57) for the projective curves, already bring to light their importance in the study of surfaces of S_3 over the group of all projectivities. This importance will be confirmed by the further properties which we obtain in this and the following sections.

We begin with some observations, which are interesting from the historical viewpoint (see the Notes following § 33, p. 69), as for the use to which they will be put in connecting the considerations of the present Part with those of the previous Part, and in assisting with the algorithmic developments in which we shall be engaged.

We consider a non-developable surface, F, which could now be a quadric, and we conserve for it the notation of § 26, p. 53. Let L be a curve of F, which is not an asymptotic $u = \text{const}$. The asymptotic tangents to F — relative to the system $u = \text{const}$. — taken at the points of L, then generate a non-developable ruled surface, R, which contains L. Employing this surface we can, from the remarks towards the end of § 23, p. 48, therefore give a first intrinsic definition of the cross-ratio of four points of L, or rather we can introduce on L a projective parameter, t (defined to within a linear fractional substitution, which is reversible and has constant coefficients).

With the aim of determining such a parameter t, we take — as is allowable — the equation of L on F in the form

$$v = v(u) . \tag{19}$$

In this case, the generator of R passing through the point x of L is the line joining x and x_v. Hence, with the obvious notation, this line has

PLÜCKER coordinates r_{ik} given by the six equations represented symbolically by

$$r(u) = (x, x_v),$$

where in the right-hand sides v is expressed as the function of u given by (19).

Whence, by differentiating totally with respect to u, and by using (1) and (11), we have

$$\frac{dr}{du} = (x_u, x_v) + (x, x_{uv}) + \gamma \frac{dv}{du}(x, x_u);$$

so that, recalling (3) and (11), we obtain

$$\frac{dr_{12}}{du}\frac{dr_{34}}{du} + \frac{dr_{13}}{du}\frac{dr_{42}}{du} + \frac{dr_{14}}{du}\frac{dr_{23}}{du} = \frac{1}{2}.$$

Similarly, differentiating again and employing (1), (9) and (11), we have

$$\frac{d^2r_{12}}{du^2}\frac{d^2r_{34}}{du^2} + \frac{d^2r_{13}}{du^2}\frac{d^2r_{42}}{du^2} + \frac{d^2r_{14}}{du^2}\frac{d^2r_{23}}{du^2} = -\frac{1}{2}\gamma^2\left(\frac{dv}{du}\right)^4 - 2\beta\gamma\frac{dv}{du} - q_{11}.$$

It is therefore sufficient to apply the results of E. CARTAN [27], in order to conclude that the various parameters t are the integrals of the differential equation

$$\Lambda(t, u) + \frac{1}{2}\gamma^2\left(\frac{dv}{du}\right)^4 + 2\beta\gamma\frac{dv}{du} + q_{11} = 0, \tag{20}$$

where $\Lambda(t, u)$ stands for the Schwarzian of t with respect to u:

$$\Lambda(t, u) = \frac{\dfrac{d^3t}{du^3}}{\dfrac{dt}{du}} - \frac{3}{2}\left(\frac{\dfrac{d^2t}{du^2}}{\dfrac{dt}{du}}\right)^2.$$

In a perfectly analogous manner, if L is not an asymptotic $v = \text{const.}$, we can consider the non-developable ruled surface generated by the asymptotic tangents to F — relative to the system $v = \text{const.}$ — at the separate points of L. This gives a second intrinsic definition of the cross-ratio of four points of L, or rather allows the introduction on L of a new projective parameter, τ. This parameter is given by the equation

$$\Lambda(\tau, v) + \frac{1}{2}\beta^2\left(\frac{du}{dv}\right)^4 + 2\beta\gamma\frac{du}{dv} + q_{22} = 0, \tag{21}$$

which is the analogue of (20), where we now naturally express u as the function of v which is obtained by inverting (19).

We propose first to prove that:

On a non-developable surface F of S_3, the projective curves possess the characteristic property that the two preceding definitions of cross-ratio — applied to any set of four points of such a curve — always give identical values. In other words, for the projective curves, and for only those curves, *the two intrinsic parameters t and τ defined above are linear fractional functions of each other.*

Those curves which satisfy the condition indicated for the parameters, are those for which, from (19)—(21),

$$\Lambda(t, \tau) = 0$$

holds. Consequently, from those equations, this relation is equivalent to

$$\begin{aligned} &2\,du\,(dv)^2 d^3u - 3\,(dv)^2 (d^2u)^2 + 3\,(du)^2 (d^2v)^2 - 2\,(du)^2 dv\, d^3v + \\ &\quad + \beta^2 (du)^6 - 2\,q_{11} (du)^4 (dv)^2 + 2\,q_{22} (du)^2 (dv)^4 - \gamma^2 (dv)^6 = 0\,. \end{aligned} \tag{22}$$

It now suffices to apply the results of § 26, pp. 53—54, in order to see that (22) can also be written in the condensed form

$$(x\,dx\,d^2x\,d^3x) + (\xi\,d\xi\,d^2\xi\,d^3\xi) = 0\,,$$

which, from § 27, p. 57, is the differential equation of the projective curves.

The assertion is therefore proved. Further, from (22), using the abbreviations (5), we see that:

The differential equation of the projective curves can also be written in the simpler form

$$2\,w\,w'' - 3\,w'^2 + \beta^2 w^6 - 2\,q_{11} w^4 + 2\,q_{22} w^2 - \gamma^2 = 0\,. \tag{23}$$

More generally, recalling the formulae of § 26, pp. 53—54, we can change the equation (18) into

$$\begin{aligned} &(1-k)\,(2\,w\,w'' - 3\,w'^2 + \beta^2 w^6 - 2\,q_{11} w^4 + 2\,q_{22} w^2 - \gamma^2) + \\ &+ 2\,(1+k)\,[2\,(\beta\,w^3 - \gamma)\,w' - \beta_u w^5 - 2\,\beta_v w^4 + 2\,\gamma_u w^2 + \gamma_v w] = 0\,, \end{aligned} \tag{24}$$

which is therefore a simplified form of *the differential equation of the principal lines of a given index k.* Equation (24), as is *a priori* evident from § 27, p. 56, and as immediately ascertained directly, reduces to (4) and (23) for the values $k = 1$ and $k = -1$ respectively.

§ 29. The use of the Laplace invariants and of the infinitesimal invariants. — We have already found (§§ 27, 28, pp. 57, 58) two differential characteristic properties of the single projective curves. We can further — following TERRACINI (see the Notes which follow § 33, p. 69) — assign a necessary and sufficient condition to an ∞^1 family of curves, which ensures that it consists of projective curves. With this end in view, we consider on a non-developable surface of S_3 any ∞^1 family of curves. Such a family determines, uniquely, a c o n j u g a t e family, such that the point or tangent coordinates of the given surface — referred to the two families as curvilinear coordinates — satisfy respectively one or other of two LAPLACE equations. Then:

Any of the curves of the given family is a projective curve if, and only if, the sum of the second invariants of the two LAPLACE *equations, just considered, is identically zero.*

To the geometric elements previously introduced we can add, by means of local considerations, a further type due to BOMPIANI (cf. the Notes referred to above); accordingly, we now pass on to make this observation more specific.

Let r_1, r_2 and r_3 be three lines of a pencil with centre O, and let L denote a curve in the plane of this pencil with a simple point at O. If P is a point of L, different from O, then we put $r = OP$ and we consider the cross-ratio $(r_1 r_2 r_3 r)$. It can be shown that the principal value of this cross-ratio as P tends to O (that is, the infinitesimal of the first order obtained by subtracting unity from the ratio between the cross-ratio itself and its limiting value) does not depend on r_3, and is therefore an *infinitesimal projective invariant* of r_1, r_2 and of the second order element of L with centre O.

We can, in particular, use this result with reference to a surface F, for which we retain the hypotheses and the notation of § 26, p. 53. If O is a general simple point of F, then it is well-known that the section of F by the tangent plane at O has a double point at O, and that the two branches of this section at O touch the asymptotic directions. If we take L as the branch touching $du = 0$, and r_1 and r_2 respectively as the tangent lines at O to the curves characterized by a given ratio $du{:}dv$ and by $dv = 0$, then the above invariant, multiplied by a numerical factor -3, is equal to $\gamma \frac{dv^2}{du}$. Similarly, by interchanging the roles of u and v, we obtain a meaning for $\beta \frac{du^2}{dv}$. We thus have geometric interpretations for the *elementary forms* $\beta \frac{du^2}{dv}$ and $\gamma \frac{dv^2}{du}$ — as introduced by BOMPIANI — the sum of which gives the *linear-projective element* of FUBINI (§ 26, p. 54).

We now consider four lines r_1, r_2, r_3 and r_4, lying in an S_3 and passing through a point O, which are subject to the one condition that the first of them should not be coplanar with any two of the other three. If L is a curve passing simply through O and touching r_4 at that point, and if P denotes a point of L different from O, then the cross-ratio of the planes joining r_2, r_3, r_4 and the point P to r_1 assumes — as P tends to O — a principal value, which is an *infinitesimal projective invariant* of r_1, r_2, r_3 and of the second order element of L with centre O. For the particular case in which L is a curve lying on a surface F, and r_2 and r_3 are the asymptotic tangents at O, it can be shown that this invariant is unchanged for the various positions of r_1; it now depends only on F and the first order element of L with centre O, and so can be determined by a given value of (du, dv). It is then equal to precisely $-\beta \gamma \, du \, dv$, which gives a geometric interpretation for the *normal quadratic differential form of F* (§ 26, p. 54).

Again, let L be a curve of F, for which O is a simple point with coordinates $x = (x_1, x_2, x_3, x_4)$. If P is a point of L, different from O, and sufficiently near to O, then each of the asymptotics $du = 0$, $dv = 0$ passing through P — in the neighbourhood of O — will meet the asymptotics of the other system, passing through O, in a point. Let P_1 and P_2 respectively be the two points so obtained. We now denote by ω the osculating plane to L at O, and by r any line of ω passing through O. The principal value of the cross-ratio determined by ω and the three planes joining P_1, P, P_2 to r does not depend on r, and is an *infinitesimal projective invariant* of L and F at O. It is given by

$$-\frac{1}{12}\frac{(x\,dx\,d^2x\,d^3x)}{(x\,x_u\,x_v\,x_{uv})}\frac{1}{(du\,dv)^2}\,.$$

Using (6) and (18), we see immediately from the above that:

The ratio of this last infinitesimal invariant to its dual, coincides with the finite invariant k of § 27, p. 56, which — as is seen there — permits a characterization of the pangeodesics and the projective curves by means of the conditions $k = 1$ and $k = -1$ respectively.

§ 30. Some classes of surfaces on which the concept of cross-ratio is particularly simple. — On the surface F (cf. § 28, p. 57), in order that the asymptotics $u =$ const. should be principal curves, it is necessary and sufficient that (24) should be satisfied by $w = 0$, which gives $\gamma = 0$. From the second relation of (1), this occurs if, and only if, the above asymptotics are rectilinear. Hence:

A family of asymptotics of a non-developable surface of S_3 consists of principal curves if, and only if, it consists of lines.

On this hypothesis, the ruled surface which is generated by the asymptotic tangents — with respect to the other system — at the points of one of those lines, is well-known to be a regulus. Thus the notion of the cross-ratio of four points of such a line — arising from § 28, p. 57 — coincides with the usual notion.

For any surface F (possibly not ruled) we can obtain the intrinsic parameter, τ (cf. § 28), relative to an asymptotic $u =$ const., by integrating the differential equation

$$\Lambda(\tau, v) + q_{22} = 0\,,$$

which arises from (21) by giving u a constant value. Hence the parameter τ does not depend on that constant if, and only if, q_{22} is independent of u. Recalling the second equation of (12), we then have:

In order that the asymptotics $u = \text{const.}$ should be set in point-point correspondence, with conservation of cross-ratio, by the asymptotics of the other system, it is necessary and sufficient that F should satisfy the condition

$$2\,\gamma\,\beta_v + \beta\,\gamma_v = 0\,. \tag{25}$$

Since (25) holds for $\beta = 0$ and for $\gamma = 0$, it follows that:

On any non-developable ruled surface of S_3 the asymptotics of either of the two systems are set in point-point correspondence with conservation of cross-ratio by the curves of the other system.

In the case of a quadric this theorem is classical. Moreover, according to a previous observation, one half of the theorem coincides with a well known result of P. SERRET which states that the curvilinear asymptotics induce a projective correspondence between any two generators of a ruled surface.

We now propose to determine the totality of surfaces, F, possessing the above property relative to both systems of asymptotics. For such an F, besides (25), we will have similarly the relation

$$2\,\beta\,\gamma_u + \gamma\,\beta_u = 0\,. \tag{26}$$

We exclude the ruled surfaces, for which the result is already acquired, that is, we suppose $\beta\gamma \neq 0$. Then the most general solution of (25) and (26) is given by

$$\beta = \frac{U^2}{V}\,,\quad \gamma = \frac{V^2}{U}\,,$$

where U and V are, respectively, arbitrary (non-zero) functions of u and v only. It then suffices to replace the asymptotic coordinates u, v by $\int U\,du$ and $\int V\,dv$ respectively, in order to reduce to the case in which

$$\beta = \gamma = 1\,. \tag{27}$$

Whence the conditions (12) for the integrability of (1) reduce to

$$q_{11\,v} = 0,\quad q_{22\,u} = 0,\quad q_{22\,v} = q_{11\,u}\,,$$

which gives

$$q_{11} = 2(c\,u + h),\quad q_{22} = 2(c\,v + k)\,,$$

where c, h and k denote three arbitrary constants. From (11) and (1), we now see that the surface F is obtained by integrating the system

$$x_{uu} = x_v + (c\,u + h)\,x,\quad x_{vv} = x_u + (c\,v + k)\,x\,. \tag{28}$$

Such surfaces are the well-known surfaces of coincidence (that is, the surfaces for which the canonical lines coincide at every point: cf. FUBINI-ČECH [48], p. 157); and in this class of surfaces are all the tetrahedral surfaces

$$x_1^{a_1}\,x_2^{a_2}\,x_3^{a_3}\,x_4^{a_4} = \text{const.}\qquad (\text{where } a_1 + a_2 + a_3 + a_4 = 0)\,.$$

Consequently:

The surfaces for which the two systems of asymptotics induce on each other a point-point correspondence preserving the cross-ratios, are precisely the ruled surfaces and the surfaces of coincidence.

On any such surface four asymptotics of the same system intrinsically define a cross-ratio, which is the constant cross-ratio of the four points in which they cut the members of the other system of asymptotics. Hence, with any four points of a surface of this type, we can associate *two different cross-ratios*, given by the cross-ratios relative to the sets of four asymptotics of the two systems which pass through the points. In the case of a ruled surface, which is neither developable nor a quadric, the cross-ratio defined by the four rectilinear asymptotics manifestly coincides with the cross-ratio in the sense of E. CARTAN (cf. the last but one paragraph of p. 52). On the quadrics of S_3, the two cross-ratios are quite elementary and have been investigated by STUDY [151] by means of an interesting analytic treatment; cf. also WEISS [170].

If, for a surface F, only one of (25), (26) holds, then there is similarly defined *just one cross-ratio* for any four points of F. We now propose to determine completely such a class of surfaces.

We suppose, for instance, that (25) holds, which is tantamount to admitting that *any two asymptotics $u = const.$ are set in one-one correspondence with conservation of cross-ratios by the asymptotics $v = const.$* This is then equivalent to saying that $\beta^2\gamma$ is a function of u only, and this function is non-zero since we shall exclude the ruled surfaces. We can therefore replace the parameter u by the function $\int \sqrt[3]{\beta^2\gamma}\, du$ of u only. This has the effect of reducing the relation between β and γ to the form

$$\gamma = \beta^{-2}. \tag{29}$$

Then, using (25), we see that the second equation of (12) is reduced to $q_{22u} = 0$; whence q_{22} is a function of v only. Then, by choosing a suitable function of v to replace the parameter v, we can ensure that we have simply

$$q_{22} = 0 \tag{30}$$

(at the same time, in order to retain the validity of the first two equations (11), we must also change the x_i's by a common factor which is a suitable function of v).

Taking account of (29), we can satisfy the first of (12), most generally, by assuming

$$q_{11} = 3\alpha_u, \quad \beta = \alpha_v^{-1}, \tag{31}$$

where α denotes any function of u, v which is not independent of v. This function must be chosen in such a way that the two remaining equations of (12) are satisfied. Now, the first of these is in fact satisfied from (29) and (30); and the second becomes, from (29) - (31),

$$(\alpha_v^2)_{uuu} - (\alpha_v^{-1})_{vvv} = 3\,\alpha_v^2\,\alpha_{uu} + 12\,\alpha_u\,\alpha_v\,\alpha_{uv}\,.$$

In conclusion, we have:

To any solution $\alpha = \alpha(u,v)$ (not independent of v) of this partial differential equation of the fourth order there corresponds a surface of the required type, which can be obtained by solving the (completely integrable) system (1), *when the coefficients are expressed by means of* (11), (29), (30) *and* (31).

§ 31. Point correspondences which conserve the projective curves. — We now propose to prove the following theorem, which will justify the name projective curves (introduced in § 27, p. 57).

In order that a point correspondence between two non-developable surfaces of S_3 should be projective (that is, subordinate to a homography or correlation of S_3), it is necessary and sufficient that it should transform the projective curves of one into the projective curves of the other. There are, besides, some exceptions to the sufficiency of this condition, but only in the case of those surfaces (depending on two essential constants) which admit an ∞^2 group of self-collineations and which have the normal form φ_2 with zero curvature.

Since the necessity of the enunciated condition has already been proved (in § 27, p. 57), it suffices to consider only the sufficiency. Let F and $\overline{F}$ be two non-developable surfaces of S_3, for which we adopt the notation of the preceding sections, and an analogous notation with bars above the symbols, respectively. We suppose that there is a biregular correspondence, T, between F and $\overline{F}$ which transforms the projective curves of F into the projective curves of $\overline{F}$. In view of (22) and ($\overline{22}$), on F and $\overline{F}$ *the asymptotic curves are the singular curves of the differential equation of the projective curves.* Hence it follows that the correspondence T transforms the asymptotics of F into the asymptotics of $\overline{F}$. Thus T can be represented analytically so that homologous points of F and $\overline{F}$ are given by the same values of suitable asymptotic parameters u and v.

With such a choice of parameters, the differential equations (22) of the projective curves of F will have the same integrals as (and can therefore be identified with) the analogous equations relative to $\overline{F}$. This implies that

$$\beta^2 = \bar{\beta}^2, \quad \gamma^2 = \bar{\gamma}^2, \tag{32}$$

and also

$$q_{11} = \bar{q}_{11}, \quad q_{22} = \bar{q}_{22}\,. \tag{33}$$

From (32), we distinguish between the following alternatives. If

$$\beta = \bar{\beta},\ \gamma = \bar{\gamma} \quad \text{or} \quad \beta = -\bar{\beta},\ \gamma = -\bar{\gamma},$$

it is well known — and follows immediately from (1), (7), (9), (11) and (33) — that T is subordinate on F and $\overline{F}$ to a homography and a correlation, respectively, of the ambient space. It remains to examine only the possibilities arising from

$$\beta = \bar{\beta}, \quad \gamma = -\bar{\gamma}, \tag{34}$$

or else

$$\beta = -\bar{\beta}, \quad \gamma = \bar{\gamma}.$$

It will suffice to consider the first of these cases, as the second reduces to it if we interchange the roles of u and v.

Subtracting from the first two relations of (12) the corresponding relations of ($\overline{12}$), we obtain (25) and (26). Thus we can again, as in § 30, p. 62, suppose that we have chosen u and v such that (27) holds. Then, using (33), (34) and (27), we see that (12) and ($\overline{12}$) give rise to

$$q_{11\,v} = 0, \quad q_{22\,u} = 0, \quad q_{22\,v} = q_{11\,u}, \quad q_{22\,v} = -q_{11\,u},$$

so that q_{11} and q_{22} must be constant. We shall denote them by $2h$ and $2k$ respectively.

Taking into account (11), (27) and (34), we now see that the equations (1) and the analogous equations ($\overline{1}$) assume the form

$$\left.\begin{aligned} x_{uu} &= x_v + h\,x \\ x_{vv} &= x_u + k\,x \end{aligned}\right\} \quad (35) \qquad \left.\begin{aligned} x_{uu} &= x_v + h\,x \\ x_{vv} &= -x_u + k\,x \end{aligned}\right\} \quad (\overline{35})$$

where h and k are constants. F and $\overline{F}$ are thus surfaces of coincidence, the first of the above systems being obtained from (28) by putting $c = 0$. More precisely, we are now dealing with surfaces of coincidence characterized by the property of possessing an ∞^2 group of self-collineations; and a second method of characterizing them, is by adding to this last the further property, which is immediately deduced from (35) and ($\overline{35}$), that they possess normal forms φ_2 with zero curvature (cf. Fubini-Čech [48], pp. 157—158 and pp. 389—391). Since the correspondence, T, defined — by equating the asymptotic parameters — between the surfaces F, $\overline{F}$ which are integrals of the systems (35) and ($\overline{35}$) is manifestly non-projective, the theorem is proved.

We add that, for general values of h, k, the surfaces F and $\overline{F}$ are tetrahedral surfaces which are projectively identical (see Fubini-Čech, loc. cit.). This does not contradict what has been previously said about T, when we remark that the surface which is an integral of any system (35) admits an ∞^2 *group of transformations preserving its projective curves*. These transformations can be represented by

$$u' \pm u = \text{const.}, \quad v' \pm v = \text{const.}, \tag{36}$$

and possess the following properties. In (36) we can choose both signs positive, both signs negative, or one sign positive and the other negative; the corresponding self-transformation is then respectively homographic, or correlative, or non-projective. Any two transformations of the last type are not even projective deformations, but have for their product a projective transformation.

§ 32. Point correspondences which preserve the principal lines. — We now suppose that we have a point transformation, T, between the two surfaces F and $\overline{F}$ — which are neither developable nor quadrics — of S_3, such that the principal lines of each of them are transformed into the principal lines of the other. We proceed as in § 31, p. 64, and we adopt the notation of that section. It will be seen that T can be taken as defined by the property of associating the points of F and $\overline{F}$ which have the same asymptotic coordinates u, v. This follows from the differential equation of the ∞^4 system formed by the principal lines of F; this equation can be deduced from (24) by solving for $(1+k)/(1-k)$ and equating to zero the total derivative with respect to v of the expression so obtained. The differential equation is therefore

$$\begin{aligned} &w\, w''' \left[2(\beta w^3 - \gamma)\, w' - \beta_u w^5 - 2\, \beta_v w^4 + 2\, \gamma_u w^2 + \gamma_v w\right] - \\ &- w'' \left[\beta^3 w^9 - (2\, \beta\, q_{11} + \beta_{uu})\, w^7 - (2\, \gamma\, q_{22} - \gamma_{vv})\, w^2 + \gamma^3\right] + \cdots = 0, \end{aligned} \tag{37}$$

where only the terms of the highest order of differentials and some of those with lower order are specified. The differential equation (37) has the asymptotics (and pangeodesics) of F as singular curves, and so the above assertion regarding T follows.

The hypothesis made on T is equivalent to admitting that the equation $(\overline{37})$, relative to $\overline{F}$ and analogous to (37), must coincide with (37); hence, with other relations, we now obtain

$$\frac{\beta}{\bar\beta} = \frac{\beta^3}{\bar\beta^3} = \frac{\gamma}{\bar\gamma} = \frac{\gamma^3}{\bar\gamma^3} = \frac{2\,\beta\, q_{11} + \beta_{uu}}{2\,\bar\beta\, \bar q_{11} + \bar\beta_{uu}} = \frac{2\,\gamma\, q_{22} - \gamma_{vv}}{2\,\bar\gamma\, \bar q_{22} - \bar\gamma_{vv}}.$$

If $\beta\gamma \neq 0$ (and therefore $\bar\beta\bar\gamma \neq 0$), then, from the equality of the first four ratios, only the relations

$$\beta = \bar\beta,\ \gamma = \bar\gamma, \quad \text{or} \quad \beta = -\bar\beta,\ \gamma = -\bar\gamma$$

appear. Hence, from the remaining two fractions, we have in either case

$$q_{11} = \bar q_{11}, \quad q_{22} = \bar q_{22}.$$

The same conclusion is reached likewise if $\beta\gamma = 0$, since then not both of β, γ are zero, as F is not a quadric; but in this case we must utilize, for this purpose, some of the terms of (37) denoted by the dots.

By employing (1), (7), (8), (11) and § 27, p. 57, we see immediately from the above results that:

The only point correspondences between two surfaces (which are neither developables nor quadrics of S_3), which transform the principal curves of one into the principal curves of the other, are those induced between the surfaces by the homographies and correlations of space. Moreover, the homographic correspondences preserve the index of each principal curve, and the correlations change the index into its reciprocal.

Let F be a surface of S_3, which is neither developable nor a quadric. Through four points A, B, C and D, chosen arbitrarily in a sufficiently restricted region of F, there passes one, and only one, principal curve, which we shall denote by L. The index k of L — as defined in § 27, p. 56 — is thus a projective invariant of the set of four points; we may therefore call it the *cross-ratio of the points A, B, C, D on F*. Adopting for a moment this nomenclature, we see from the last theorem that:

The homographic correspondences are the only point correspondences, between surfaces — neither developables nor quadrics — of ordinary space, which preserve cross-ratios.

From this result, one would be inclined to adopt definitely the above number k as the cross-ratio $(ABCD)_F$. But such a definition presents an inconvenient consequence, since k is a *symmetric* function of the four points, whilst it would be desirable that the customary relations should connect the various cross-ratios obtained from the different ways of ordering the points. This last condition we can satisfy by taking $(ABCD)_F$ as the *cross-ratio of the points A, B, C, D on one or other of the two ruled surfaces asymptotically circumscribed to F along L*. In this way, we obtain *two values for the cross-ratio*: and this conforms with what has already been accomplished in other circumstances (§ 30, p. 63), the process being precisely the same in the particular case of a quadric (but this case has been excluded in the present section).

If we wish to obtain only one value for the cross-ratio, then we should take $(ABCD)_F$ equal to the cross-ratio of A, B, C, D on the curve L, calculated by the method of § 25, p. 50. However, this definition is far more complicated than the previous one, and does not allow extension to the quadric surfaces; further, it presents an artificial discontinuity, relative to sets of four points of F lying on a possible "plane cone curve" of F (cf. the following section), in so far as the cross-ratio on L is determined by totally different methods in the two cases in which L is or is not planar.

Other univalent definitions which are possible, but rather involved, for the cross-ratio $(ABCD)_F$, are obtained by taking it equal to the cross-ratio of A, B, C, D — in the sense of E. CARTAN [37] — on the ruled surface (non-developable in general) which is generated by the projective normals, or by the edges of GREEN, or by some other canonical line, relative to the separate points of F lying on L. The comparison of the various definitions suggests the problem of determining those surfaces for which two different definitions result in the same value for the cross-ratio.

§ 33. On the plane cone curves of a surface. — If L is a principal curve of a surface F (which is neither developable nor a quadric), then there is attached to it a constant which is generally uniquely determined by L. This constant we have called the index of L (§ 27, p. 56), and is that number k for which L satisfies the corresponding equation (18) or (24). In virtue of the linearity of these equations in k, L is exceptional if, and only if, it gives rise to two distinct values of k and, in this case, the index of L is indeterminate. From § 27, pp. 56—57, we see, in particular, that:

If a curve, lying on a surface (neither developable nor a quadric) of S_3, is a member of any two of the four classes of curves: plane sections, cone curves, pangeodesics, projective curves, then it is also a member of the other two classes.

Every such curve, relative to a surface F, will be called briefly a *plane cone curve* of F. The investigation of the plane cone curves of a given surface F is equivalent to *the determination of the integrals common to the differential equations* (4) *and* (23), which characterize, respectively, the pangeodesics and the projective curves of F (§§ 26, 28, pp. 54, 59). If we add to these two equations a third, which we obtain from (4) by differentiating totally with respect to v, then it is easily seen that we can eliminate w' and w'' from these equations and so obtain an algebraic equation in w:

$$\beta^5 w^{15} + \cdots + \gamma^5 = 0\,, \tag{38}$$

in which we have specified only the extreme terms; (38) is not identically satisfied since — as F is not a quadric — β and γ are not both zero. Recalling the significance of w from the first relation of (5), we conclude that:

Whilst on a quadric of S_3 every plane section is a cone curve, for no other non-developable surface of S_3 can there be more than a simple infinity of plane cone curves. Through the general point of such a surface, there never pass more than 15 *plane cone curves of the surface.*

We could, in fact, diminish the number 15 given here by establishing — as an algebraic differential consequence of (4) and (38) — an algebraic equation in w of degree less than 15. Here we shall not deepen the investigation, which becomes rather complicated from the formal viewpoint; one ought to consider in more detail the conditions of compatibility of the differential equations (4) and (23), the coefficients of which — it should be remembered — are connected by relations (12).

We merely add that the surfaces which contain only one ∞^1 family of plane cone curves (that is, one family of plane pangeodesics, of which only one passes through the general point) have been all obtained by FUBINI. They are representable analytically in finite terms, and *depend on six arbitrary functions of an argument* (and not on seven as is erroneously stated in FUBINI-ČECH [48], pp. 539—540, where the necessity — for the four curves k_i considered on the developable T — to cut projective quadruples on the different generators seems to have been omitted). Also surfaces are known, which *depend on two arbitrary functions* and which contain two distinct ∞^1 families of plane cone curves: for an extensive study, and for the bibliographical indications on them, cf. B. SEGRE [93]. Such surfaces include, among others, the *tetrahedral surfaces* (already encountered from a different angle in §§ 30, 31, pp. 62, 65), each of which — as is easily verified — contains six ∞^1 families of plane cone curves.

Historical Notes and Bibliography

The projective curves of a surface appear for the first time in B. SEGRE [95], and there they are defined by means of the property expressed in the first theorem of § 28, p. 58; the analytic representation is also deduced, as is the proposition of § 31, p. 64. Later, TERRACINI [155] obtained, for the systems of ∞^4 projective lines, the characterization which here appears at the beginning of § 29, p. 60, together with another, which is less significant. B. SEGRE [97] afterwards established the property for projective curves, which we here use for their definition (§ 27, p. 57), based on the consideration of the invariant I. And this has enabled him to introduce at the same time the ∞^4 curves which we have called principal curves. BOMPIANI [22] also noted these curves a little later, by a different method; to him we owe the results of § 29, p. 61 which follow the proposition of TERRACINI.

The various results of §§ 30, 32, 33, pp. 61—64, 66—69, appear here for the first time, apart from the definition given in the penultimate paragraph of § 32, which is already to be found in BOMPIANI [23, 24].

Part Five

Some Differential Properties in the Large of Algebraic Curves, their Intersections, and Self-correspondences

We shall show in the present Part how the *theory of residues of analytic functions* can be employed in the — local or global — study of certain *differential properties of analytic and algebraic curves*; these properties will concern differential elements defined by intersections, or by fixed points of analytic or algebraic correspondences. Further, with regard to the above theory, intrinsic geometric interpretations will be introduced, which will imply geometric meanings for the differential invariants as they arise.

§ 34. The residues of correspondences on curves, and a topological invariant of intersection of two curves on a surface which contains two privileged pencils of curves. — We now propose to extend the considerations of § 8, p. 15 (Part Two), relative to the residues of a correspondence, in the simplest case of curves ($n = 1$). Let O be a united point of an analytic self-correspondence T of an analytic curve. If x is an allowable complex coordinate on the curve in the neighbourhood of O, at which point x takes the value x_0, then T is represented, in such a neigbourhood of O, by an equation of the type

$$X = f(x) .$$

Now, if O is a simple united point of T (such that $f'(x_0) \neq 1$), the residue ω of T at O is given by (§ 8, p. 15)

$$\omega = \frac{1}{1 - f'(x_0)} .$$

More generally, writing the equation of T, in the neighbourhood of O, in the implicit form

$$\Theta(x, X) = 0 , \tag{1}$$

we have $(\Theta_X + \Theta_x)_0 \neq 0$, in virtue of the simplicity of the united point O, and, moreover,

$$\omega = \left(\frac{\Theta_X}{\Theta_X + \Theta_x}\right)_0 .$$

Therefore:

The residue ω of T at O is equal to the residue, at that point, of the analytic function

$$\varphi(x) = \left[\frac{\partial}{\partial X} \log \Theta(x, X)\right]_{X = x} . \tag{2}$$

We shall now take the above property as the definition of the residue ω of T at O, even on the hypothesis that O is not a simple united point; in this case the validity of the above expression for the residue is conserved, and it can be calculated explicitly as will be seen shortly. It is immediately deduced that ω is unaltered by any change of the parameter x (applying simultaneously the same transformation to X); and it is also unchanged if Θ is altered by an arbitrary analytic functional factor of x, X, which is non-zero for $x = x_0$, $X = x_0$. Thus *the residue ω is a topological invariant of T at O.*

From the above definition we obtain, without difficulty, the following differential expression for ω. Let n ($\geqq 1$) be the *multiplicity* of the united point O of the correspondence T given by (1); this is equivalent to saying that $(\Omega^i \Theta)_0 = 0$ for $i = 1, 2, \ldots, n-1$ (if $n > 1$), but $(\Omega^n \Theta)_0 \neq 0$, where Ω denotes the operator

$$\Omega = \frac{\partial}{\partial x} + \frac{\partial}{\partial X} ,$$

and further — as in the formula following (1) — the suffix 0 of a function denotes that it is to be calculated for $x = x_0$, $X = x_0$, so that

$\Theta_0 = 0$. Then the residue ω of T at O is given by

$$\omega = \left(\frac{n!}{(\Omega^n \Theta)_0}\right)^n \begin{vmatrix} \frac{1}{(n-1)!}\left(\frac{\partial}{\partial X}\,\Omega^{n-1}\Theta\right)_0 & \frac{1}{(n-2)!}\left(\frac{\partial}{\partial X}\,\Omega^{n-2}\Theta\right)_0 & \cdots & \left(\frac{\partial}{\partial X}\,\Theta\right)_0 \\ \frac{1}{(n+1)!}(\Omega^{n+1}\Theta)_0 & \frac{1}{n!}(\Omega^n\Theta)_0 & \cdots & \frac{1}{2!}(\Omega^2\Theta)_0 \\ \frac{1}{(n+2)!}(\Omega^{n+2}\Theta)_0 & \frac{1}{(n+1)!}(\Omega^{n+1}\Theta)_0 & \cdots & \frac{1}{3!}(\Omega^3\Theta)_0 \\ \cdots & \cdots & \cdots & \cdots \\ \frac{1}{(2n-1)!}(\Omega^{2n-1}\Theta)_0 & \frac{1}{(2n-2)!}(\Omega^{2n-2}\Theta)_0 & \cdots & \frac{1}{n!}(\Omega^n\Theta)_0 \end{vmatrix}. \tag{3}$$

If ω^* denotes the residue at O of T^{-1}, the inverse of the correspondence T, then, in view of (2) and the above conditions for Θ at O, we have at once

$$\omega + \omega^* = \text{residue at } O \text{ of } (\Omega \log \Theta)_{X=x} = n\,. \tag{4}$$

Therefore:

The sum of the residues of any analytic self-correspondence of an analytic curve and of the inverse correspondence, taken at the same united point, is always equal to the multiplicity of that united point.

With the further hypothesis that T has an involutory character (i. e., if T and T^{-1} coincide), we have manifestly $\omega^* = \omega$; thus (4) gives $\omega = n/2$. Therefore

The residue of an involutory correspondence at a united point with multiplicity n is equal to precisely $n/2$.

We remark that the preceding notions can be given a wider range if the conditions of the analyticity of T, and of the curve on which it operates, are relaxed. We can, in fact, take the expression (3) as the definition of the value of the residue ω of T at O; this is allowable if we admit for Θ suitable conditions for differentiability. By a direct calculation, less simple than that just indicated, but still not presenting difficulties, we can establish — with the present more general hypotheses — the invariance of ω, and also prove the equality $\omega + \omega^* = n$. Thus, for example, we see from (3) that the sum $\omega + \omega^*$ is given by the right-hand side of (3) where the first row of the determinant is replaced by

$$\frac{1}{(n-1)!}(\Omega^n\,\Theta)_0,\ \frac{1}{(n-2)!}(\Omega^{n-1}\,\Theta)_0,\ \ldots,\ (\Omega\,\Theta)_0\,.$$

The new determinant then has only zero elements above the principal diagonal, in view of the conditions satisfied by Θ at O. The value of the determinant therefore reduces to its principal term $n[(\Omega^n\,\Theta)_0/n!]^n$. Hence, immediately, $\omega + \omega^* = n$. Therefore (3) defines — even in the non-analytic case — a *differential invariant of order* $2n-1$ for the correspondence (1) at a n-ple united point O.

Let us now readmit the hypotheses of analyticity as before, so that — from the foregoing remarks and the definition of the residue — we have

$$\omega = \frac{1}{2\pi i} \oint \varphi(x)\, dx, \tag{5}$$

where $\varphi(x)$ is given by (2) and the integral is taken round a contour, in the plane of the complex variable x, which encircles the point x_0 in a positive sense precisely once, and which does not pass through, or contain in its interior, any other x belonging to a united point of T. Employing (5), one proves easily the following important theorem.

Let T^* be an analytic correspondence of an analytic curve into itself depending on a continuous parameter t, such that T^* tends to T as $t \to t_0$. If T admits O as a united point with multiplicity n, then T has a certain number, h, of united points which tend to O as $t \to t_0$ $(1 \leq h \leq n)$; it is known that, if these united points are denoted by $O', O'', \ldots, O^{(h)}$ and the corresponding multiplicities by $n', n'', \ldots, n^{(h)}$, then $n = n' + n'' + \cdots + n^{(h)}$. Moreover, as $t \to t_0$, the sum of the residues of T^* at the points $O', O'', \ldots, O^{(h)}$ has a definite limit, which is precisely the residue of T at O (although the separate limits of those residues may not be finite in the case of $h > 1$).

This result allows the introduction of a geometric interpretation of the residue ω, considered as a differential invariant of the curve (1) and the line $x = X$ at the point $x = x_0$, $X = x_0$ in the (x, X) plane.

More generally, we consider an analytic surface, F, endowed with two privileged pencils of analytic curves, $\mathcal{A}$ and $\mathcal{B}$, which are unisecant in the neighbourhood of a given point O simple on F. On F we have further two analytic curves L and L' with an isolated intersection at O. L passes simply through O, but L' may have an algebroid singularity there. We suppose, moreover, that L touches neither the member of $\mathcal{A}$ nor the member of $\mathcal{B}$ which pass through O. We can now define on L the analytic correspondence, T, which — in the neighbourhood of O — transforms the general point A of L into the point B of L if the curve of $\mathcal{A}$ passing through A meets the curve of $\mathcal{B}$ passing through B on the curve L'. It is seen immediately that T has a *united point* at O, and that the multiplicity n of O in T is equal to the *intersection multiplicity* of the curves L, L' at O. The residue ω of T at O is a differential invariant at O, relative to the curves L, L' and the families of curves $\mathcal{A}$, $\mathcal{B}$; and this invariant is unchanged when the neighbourhood of O on F is subjected to any reversible analytic transformation. Briefly, we shall call ω *the residue of L and L' at O*, and we shall give a geometric interpretation of it, which will bring to light its intrinsic character.

Firstly, from the last theorem, we have:

If L' is obtained as the limit of an analytic curve L^*, and if $O', O'', \ldots, O^{(h)}$ are the common points of L, L^* falling in the neighbourhood of O, then the *sum of the residues of L and L^* at those points always admits the same limit as L^* tends in any way to L', and this limit is precisely the residue of L and L' at O.*

From this, and from the fact that we can always choose L^* such that $O', O'', \ldots, O^{(h)}$ are simple intersections of L and L^* (and are therefore n in number), it will suffice to give a geometric meaning of ω in the elementary case for a simple intersection; that is (with reference to the point O), when we have $n = 1$. On this hypothesis, L' passes simply through O without touching L at that point, so that — in the pencil of lines tangent to F at O — we obtain four distinct lines l, l', a, b, touching simply at O the curves L, L' and the curves of $\mathcal{A}$ and $\mathcal{B}$ passing through O. Denoting by λ the coefficient of dilatation of T at O, we see directly (or by using § 3, p. 6) that

$$\lambda = (a\, b\, l\, l')\ .$$

In virtue of § 8, p. 15, we therefore obtain

$$\omega = \frac{1}{1-\lambda} = (a\, l\, l'\, b)\ ,$$

which gives the required interpretation.

§ 35. A complement of the correspondence principle on algebraic curves. — In particular, we can apply the results of § 34, p. 69, to the study of *algebraic correspondences* of an algebraic curve into itself. Let T be such a correspondence, relative to any irreducible algebraic curve, C; and let O be a point of C which — considered as the centre of a fixed branch of this curve — is a united (or fixed) point of T. In the neighbourhood of O on that branch we can, as is well known, introduce a (complex) uniformizing parameter, t; t is determined to within a reversible analytic transformation, and T operates on it analytically. It follows — in view of § 34, p. 70 — that we can define *the residue ω of T at the point O* (regarded as the centre of a branch of C), and that this is a *birational invariant* of T at O (giving rise, in general, to a complex number).

(1), (2), and (5) suggest the definition of ω at points which are not united for T. We shall assume that in this case $\omega = 0$. From this and the above it follows that:

At every point of any algebraic curve, the residue of a correspondence which is the sum of two or more correspondences is always equal to the sum of the residues, at that point, for the component correspondences.

Indeed, if $T = T_1 + T_2$, and if $\Theta_1(x, X) = 0$, $\Theta_2(x, X) = 0$ are the equations of T_1 and T_2 in the neighbourhood of the point O of C, then the equation of T in this neighbourhood can be taken as

$\Theta(x, X) = 0$, where $\Theta(x, X) = \Theta_1(x, X) \cdot \Theta_2(x, X)$. The result now follows immediately from (2), (5), since

$$\left[\frac{\partial}{\partial X} \log \Theta(x, X)\right]_{X=x} = \left[\frac{\partial}{\partial X} \log \Theta_1(x, X)\right]_{X=x} + \left[\frac{\partial}{\partial X} \log \Theta_2(x, X)\right]_{X=x}$$

We propose to prove the following complement of the correspondence principle:

On an algebraic curve C with genus p, any algebraic correspondence T with indices (α, β), with valency $\gamma \gtreqless 0$, and not having the identity as component, admits at its various united points (simple or multiple) residues having the integer $\beta + \gamma p$ as sum. Symbolically:

$$\sum_{O \in C} \omega(O) = \beta + \gamma p. \tag{6}$$

We note firstly that (6) has an invariant character over the birational transformations of C, and we add further that, applying (6) to the inverse correspondence T^{-1}, we obtain

$$\sum_{O \in C} \omega^*(O) = \alpha + \gamma p; \tag{6*}$$

hence it suffices to add (6) and (6*) term by term, recalling (4), to obtain the classical correspondence principle of CAYLEY-BRILL-HURWITZ (cf. e. g. SEVERI [138], p. 233), namely:

The number of united points $= \sum_{O \in C} n(O) = \alpha + \beta + 2\gamma p$. This result thus appears as an immediate corollary of the above complement.

We begin by proving (6) for the simplest case in which T has valency $\gamma = 0$ and possesses only simple united points, O (i. e. united points with multiplicity 1). By a suitable birational transformation of C, we can always reduce C to a plane curve endowed with ordinary singularities and situated generally with respect to the reference system of coordinates. Let

$$f(x, y) = 0 \tag{7}$$

be the equation of C. Because of the hypothesis $\gamma = 0$, the correspondence T can be represented by adding to (7) a single algebraic equation (cf. SEVERI [138], p. 199):

$$\Theta(x, y; X, Y) = 0. \tag{8}$$

Now, if m is the order of C, (7) has m distinct points, P, at infinity. Further, calling b the degree of the polynomial Θ in X, Y, we shall have on C certain fixed points, Q, which we can suppose simple for C and distinct from P, so that, corresponding to the generic point (x, y) of C, (8) represents — with (X, Y) as current coordinates — a curve (of order b) cutting C in those fixed points Q, counted with certain multiplicities ν, and moreover in β variable points which are the transforms of (x, y) by T. Thence

$$mb - \sum_Q \nu = \beta\,. \tag{9}$$

Abbreviating with $F = f(X, Y)$, we consider the rational function

$$R(x, y) = \left[\frac{\dfrac{\partial(\Theta, F)}{\partial(X, Y)}}{F'_y \cdot \Theta}\right]_{\substack{X=x\\Y=y}}. \tag{10}$$

We obtain the required relation by applying a well known result (cf. for example SEVERI [126], p. 241), by which the sum of the residues of $R(x, y)$ is zero, the sum being taken over all the points (finite and infinite) of the curve (7); that is, the sum of the polar periods — divided by $2\pi i$ — of the Abelian integral $\int_c R(x, y)\, dx$ attached to this curve. From (10), we see immediately that the residue at each of the branch points of the function $y = y(x)$, implicitly defined by (7), is zero. For the remainder of the points, at which it is allowable to assume x or $1/x$ as uniformizing parameter, a simple analysis shows that:

1. at a united point O of T, the function R has a pole of the first order, with residue equal to the residue of T at the same point;

2. at a point P at infinity on C, the function R has a zero of the first order, with residue $-b$;

3. at a point Q, the function R admits a pole of the first order, with residue ν;

4. at every other point of C, the function R is holomorphic and therefore has zero residue.

We have accordingly:

$$\sum_O \omega - mb + \sum_Q \nu = 0\,.$$

It suffices now to add this equation, term by term, to (9), in order to deduce that $\sum \omega = \beta$; and this proves (6) on the present hypotheses.

The preceding analysis, with suitable modifications, furnishes the proof of (6) for any T with valency zero, endowed with united points with any multiplicities. This case can on the other hand be reduced to the above, and the result is obtained easily as follows, by considering T as a limit (as is allowable, cf. SEVERI [138], p. 266) of a variable T^* which again has valency zero, but has only simple united points. We then apply to T^* the result just established, and employ the last theorem but one of § 34, p. 72.

Before passing to the general case, we prove (6) in two other particular cases.

A general g_h^1 of C admits, as is well known, $2(h + p - 1)$ distinct double points, and defines on C an elementary correspondence, T, which associates two points — generally distinct — when they belong to the same set of points of g_h^1 itself. Such a T manifestly has indices $(h-1, h-1)$, valency $+1$, and its united points are the above $2(h+p-1)$ double points, each of which is a simple united point of T. Since — in virtue of the involutory character of T, and from § 34, p. 71 — in each of these points T has residue 1/2, it follows that (6) holds on the present hypothesis.

Secondly, we consider on C two general elementary correspondences, T_1, T_2, defined by two linear series, $g_{h_1}^1, g_{h_2}^1$, and also the product correspondence $T = T_1 \cdot T_2$. This latter has indices $((h_1-1)(h_2-1), (h_1-1)(h_2-1))$ and valency -1, and its united points are those points of pairs (O', O'') which correspond in T_1 and in T_2. The number of such pairs is known to be equal to $(h_1-1)(h_2-2)-p$ (cf. e. g. ENRIQUES-CHISINI [45], p. 74). On the other hand, in virtue of the involutory character of T_1, T_2, in the neighbourhood of any of those pairs (O', O'') we have $T^{-1} = T_1^{-1} \cdot T \cdot T_1$. Since T_1 interchanges O' and O'', it follows — also from a previous result (§ 34, p. 70) — that the sum of the residues of T in O' and O'' is equal to 1. We can thus immediately verify (6) on the present hypotheses.

Now, if T denotes any correspondence with valency $\gamma \neq 0$, we easily prove (6) by applying — as is allowable — the same equation (6) to the correspondence with valency zero which we obtain by adding to T $|\gamma|$ correspondences of the first or the second type just considered (with valencies $+1$ and -1) according as $\gamma < 0$ or $\gamma > 0$, and then using the first theorem of the present section.

§ 36. A geometric characterization of Abelian integrals and their residues. — In the preceding section we have employed the notion of an Abelian integral, and the theorem which states that the sum of the residues of any Abelian integral is zero. We now propose to expound the purely geometric content of both the notion and the theorem. The results which will appear, not only are of interest in themselves, but also the application of them in the procedure of § 35, p. 75, will lead immediately to a geometric interpretation of the residues of an algebraic correspondence (different from that of § 34, p. 72), as well as of the associated relation (6). In the sequel we shall be concerned with questions in which the residues of certain Abelian integrals repeatedly intervene; consequently, the geometric interpretations which follow can be again applied.

We first observe that every Abelian integral $\mathcal{J}$, attached to an irreducible algebraic curve C, determines on C two sets, λ, μ, of points connected by the equivalence

$$\lambda \equiv \mu + \varkappa \,, \tag{11}$$

where $\varkappa$ denotes a canonical set (effective or virtual) of C, and λ, μ are defined as follows. The set μ consists of the singular points of $\mathcal{J}$;

more precisely, a point of C at which $\mathcal{J}$ has a pole of order s, possibly superimposed on a logarithmic singularity, is counted in μ with multiplicity $s+1$; whilst a point at which $\mathcal{J}$ has only a logarithmic singularity is counted simply in μ. The set λ consists of the double points of $\mathcal{J}$, or rather of those points of C for which the equation $d\mathcal{J}/dt = 0$ holds, where t is any uniformizing parameter in the neighbourhood of the point, such that its value is finite in that neighbourhood; the number of times that such a point is counted in λ is equal to the multiplicity of the corresponding root in the above equation. The sets λ, μ have no point in common. Conversely, if two sets of points on C are given satisfying (11), then the integral $\mathcal{J}$ is defined by them to within an arbitrary non-zero *multiplicative constant factor*, c.

Let l and m be the orders of λ and μ, and let p be the genus of C; from (11) it follows that

$$l = m + 2p - 2 .$$

From the theorem on residues previously quoted, we see that, in order that $\mathcal{J}$ should have some non-zero residues, it is necessary that μ should consist of at least two distinct points; also, if μ includes only two points, the residues of $\mathcal{J}$ at those points are equal and opposite. It will thus suffice to limit ourselves to the case in which μ contains at least three distinct points, so that $m \geqq 3$. We shall then *determine geometrically the mutual ratios of the residues at these points* of any one of the integrals $c\mathcal{J}$ defined by the sets λ, μ, which, we suppose, are deprived of common points and connected by (11). Such residues will be called the residues at the points of μ relative to λ.

Putting, for brevity,

$$k = l - p = m + p - 2 ,$$

since $m \geqq 3$, we see immediately that the complete linear series $|\lambda|$ is a non-special g_l^k without neutral pairs. We can therefore substitute for C the projective image of this g_l^k, which, for simplicity, will still be called C. Hence C is now an irreducible algebraic curve, with order l in an S_k, free from multiple points; and λ becomes the set of points cut on C by a certain S'_{k-1} of S_k. Further — as a consequence of (11) — we see that μ consists of m points (distinct or infinitely near) of C, none of which lies in S'_{k-1}; the join of these points is an S_{m-2}, which coincides with S_k or is subordinate to S_k according as $p = 0$ or $p > 0$.

For greater clarity of exposition, we shall distinguish between the two alternatives in which the m points of μ are, or are not, distinct. In the first case, we easily ascertain that each set of $m-1$ of them are linearly independent, and we prove that:

The ratio of the residues of μ relative to λ in two distinct points M_1 and M_2 of C is equal to $-(M_2 M_1 N N')$, where N and N' denote respectively the *intersections of the line* M_1M_2 *with the* S_{m-3} *of* S_{m-2} *which joins the points of* μ *distinct from* M_1, M_2, *and with the above* S'_{k-1}.

The fact that the sum of the residues in the various points of μ is always zero, is then an immediate corollary of the following very simple theorem of projective geometry (in which we assume $r = m - 2$, and we take S_{r-1} as the intersection of S_{m-2} and S'_{k-1}).

Choose in an S_r, *in any manner,* $r + 2$ *points* $M_1, M_2, \ldots, M_{r+2}$ *such that any* $r + 1$ *of them are linearly independent* ($r \geq 1$), *and also an* S_{r-1} *containing none of the points; consider on each of the lines* M_iM_j ($i \neq j$, $i, j = 1, 2, \ldots, r + 2$) *the points* N_{ij} ($= N_{ji}$), N'_{ij} ($= N'_{ji}$) *in which it meets the hyperplane joining the other* r *points, and* S_{r-1}. *With each of the points* M_i *a number* $\varrho_i \neq 0$ *can then be associated, such that*

$$(M_iM_jN_{ij}N'_{ij}) = -\varrho_j/\varrho_i.$$

The $r + 2$ *numbers* ϱ *are thus determined to within a factor of proportionality* (for example, reducing S_r to an affine space by taking S_{r-1} as the hyperplane at infinity, we can identify ϱ_i with the volume of the simplex with the $r + 1$ points M, different from M_i, as vertices, taken with a suitable sign); *further, the above* ϱ_i*'s have always a zero sum.*

The case in which μ does not consist of m distinct points leads to some complications; then it is convenient to consider — for every point with multiplicity $q + 1$ — the space S_q osculating C at that point. More precisely, we have:

If μ *consists of* s (≥ 3) *points* M_i *counted with multiplicity* $q_i + 1$ ($q_i \geq 0$; $i = 1, 2, \ldots, s$; $m = \sum_{i=1}^{s} q_i + s$), *then the mutual ratios of the residues* σ_i *at the various points* M_i *of the set* μ *relative to* λ *are characterized by the relations*

$$(P_{ij}P_{ji}Q_{ij}Q'_{ij}) = -\sigma_j/\sigma_i \qquad (i \neq j;\ i, j = 1, 2, \ldots, s),$$

where P, Q *and* Q' *denote points defined in the following way.*

The two spaces S_{q_i}, S_{q_j} osculating C in the points M_i, M_j are skew, and are joined by a space which meets in precisely one point the space which joins the other spaces S_q. Then Q_{ij} ($= Q_{ji}$) denotes just this point. P_{ij}, P_{ji} are the points of the spaces S_{q_i}, S_{q_j} respectively, which are collinear with Q_{ij}. Finally, Q'_{ij} ($= Q'_{ji}$) is the point in which the line $P_{ij}P_{ji}Q_{ij}$ meets the S'_{k-1} cutting λ on C.

Choosing, in any manner, on each space S_{q_i} $q_i + 1$ points M^* which are linearly independent, we obtain in S_r $r + 2$ points with each of which is associated, as in the previous theorem, a number ϱ. The space S_{r-1} of that theorem is again taken as the S_{r-1} cut by S'_{k-1} on S_r. We can prove that:

The number σ_i *relative to* M_i *can be taken as the sum (possibly zero) of the numbers* ϱ *belonging to the* $q_i + 1$ *points* M^* *lying in* S_{q_i}. *Thus we* *obtain s numbers σ, determined — independently of the choice of the points M^* — to within a common factor of proportionality, having always a zero sum.*

§ 37. The first applications. — We shall now make some applications of the results of §§ 34, 35, pp. 69—76, to the study of pairs of algebraic curves lying on an algebraic surface, F, which contains two pencils $\{\mathcal{A}\}$ and $\{\mathcal{B}\}$ of unisecant curves. This is equivalent to saying that we suppose that F represents birationally, and without exception, the pairs of points of two given algebraic curves (namely, F is the so-called product of these curves).

Let C, C' be two algebraic curves of the above surface F, which meet, respectively, the curves $\mathcal{A}$ in α, α' points and the curves $\mathcal{B}$ in β, β' points. We suppose further that (at least) one of C, C' (C' say) is of valency zero (a condition which is always satisfied for every curve of F if either the $\mathcal{A}$'s or the $\mathcal{B}$'s are rational curves). This means that the α' curves $\mathcal{B}$ which pass through the points common to C' and one $\mathcal{A}$, vary, for the different curves $\mathcal{A}$, in a linear system: for this it is necessary and sufficient that, on F, should hold an equivalence of the type:

$$C' \equiv \sum_{h=1}^{\beta'} \mathcal{A}_h + \sum_{k=1}^{\alpha'} \mathcal{B}_k ,$$

where the $\mathcal{A}_h$'s; $\mathcal{B}_k$'s are curves of the two given pencils (cf. SEVERI [120], n. 14). Whence it is easily seen that the number of intersections $[C, C']$ satisfies

$$[C, C'] = \sum_{h=1}^{\beta'} [C, \mathcal{A}_h] + \sum_{k=1}^{\alpha'} [C, \mathcal{B}_k] = \alpha\beta' + \alpha'\beta .$$

We first suppose that the set of points common to the given curves C, C' consists of $\alpha\beta' + \alpha'\beta$ distinct points O_r (for $r = 1, 2, \ldots, \alpha\beta' + \alpha'\beta$), and we denote respectively by c_r, c'_r, a_r, b_r the four coplanar lines which touch, at O_r, C, C' and the curves $\mathcal{A}, \mathcal{B}$ which pass through this point. We then have:

The $\alpha\beta' + \alpha'\beta$ cross-ratios $(a\, c_r\, c'_r\, b_r)$ [all finite, since necessarily $a_r \neq b_r$, $c_r \neq c'_r$] have always the integer $\alpha'\beta$ as sum.

In order to establish the result, it is allowable to suppose that C is irreducible, since — if C is compounded — the assertion follows immediately by the application of the same result to the separate components of C and to C'. We define on the irreducible curve C the algebraic correspondence, T, which transforms the point A into the point B of C if the curve $\mathcal{A}$ through A meets on C' the curve $\mathcal{B}$ through B. Hence T manifestly has indices $(\alpha\beta', \alpha'\beta)$; moreover it is of valency zero, in virtue of the hypothesis assumed for C', and possesses as united points precisely the $\alpha\beta' + \alpha'\beta$ points O_r, each one being a simple united point. It now suffices to apply the theorem of § 35, p. 74 to T, and to recall the final paragraph of § 34, p. 73, in order to obtain the required result.

The general case in which C and C' have a finite number of points in common — at which they behave in an arbitrary way — is dealt with in an analogous manner, using the remarks of § 34, p. 72, which appear after the penultimate proposition. Consequently, we obtain:

Corresponding to each (isolated) intersection O of the curves C, C' of F, is defined the residue of them at O, which is a complex numerical invariant of C, C' over the birational transformations of F which are regular in a neighbourhood of O. If C or C' is of valency zero, then the sum of the residues of C, C' at their various points of intersection is always equal to the integer $\alpha'\beta$.

The results just obtained can, in particular, be applied to two algebraic curves C, C' lying on a ruled algebraic surface F. In this case, the $\mathcal{A}$'s can be taken as the generators of F, and the $\mathcal{B}$'s as the sections of F by a pencil of hyperplanes which do not have as a base point any point common to C and C'. We remark that every curve of F is then of valency zero (in the sense previously indicated), so that the foregoing results acquire validity without further restriction.

In the case, even more special, in which F is a quadric surface, we can identify the curves $\mathcal{A}$, $\mathcal{B}$ with the generators of the two reguli of F. We shall not enunciate the various properties which thus follow immediately, but instead we shall confine ourselves to giving some of them when F is metrically specialized to a spherical surface. Employing the classical formula of LAGUERRE, we obtain the following theorem.

Given two algebraic curves lying on a sphere and not touching at any common (real or imaginary) point, the sum of the cotangents of their various angles of intersection is always equal to an integral multiple of the square root of -1; it is zero if, and only if, the two given spherical curves are algebraically dependent, and this always occurs if they are both real.

Whence, by a suitable stereographic projection, we deduce that:

Given a plane algebraic curve of even order $2m$, which contains the two circular points with multiplicity m, and any other algebraic curve in its plane, which contains the two circular points with the same multiplicity $\nu \geqq 0$ (being otherwise arbitrary), such that at no finite point do the curves touch; then the sum of the cotangents of the angles between the curves at their proper common points is zero.

We add, finally, that the preceding development offers a way of seeing — without difficulty — in what way the two final results become modified on the hypothesis that the two curves behave arbitrarily at their common points.

§ 38. The equation of Jacobi, and some consequences. — In § 37, p. 79, we obtained several types of pairs of algebraic curves, lying on an algebraic surface, for which the differential elements of the two curves with centres at the various common points of the curves are linked in a rather simple manner. This suggests the study of all the relations which tie such elements; more generally, it is clear how every algebraic intersection problem might lead to a whole set of algebro-differential questions in the large, concerning the differential elements which are, in various ways, connected with that problem.

In the present section we shall deal with some questions of this type connected with the BÉZOUT theorem, that is, relative to the differential elements of the various orders determined by the intersections of two given curves in the plane. In conformity with the preceding developments, we shall often have occasion to introduce differential invariants of intersection in the form of residues of Abelian integrals; this will have four notable advantages: 1. in order to define such invariants, it will not be necessary to distinguish between the innumerable modes of behaviour of the two given curves at their common points; 2. the calculation of such invariants will be effected in the various cases *via* a regular procedure free from difficulties; 3. a geometric meaning for those invariants will be given by using § 36, pp. 76—79; 4. the theorem stating that the sum of the residues of any Abelian integral is zero, for which we obtained a simple geometric interpretation in § 36, p. 78, will furnish a property in the large — of the required type — relative to the invariants. There will be no difficulties concerning the residues of an Abelian integral, in the case where the curve to which it is attached is compounded (cf. § 37, p. 79).

If n, ν are any two positive integers, we denote by C_n and Γ_ν two algebraic curves of the plane, having orders n and ν respectively, which have no common component, and by $G_{n\nu}$ a set of $n\nu$ distinct points of the same plane. Using simple algebraic arguments, it can be proved that:

If not both of $n = 1$ or 2, $\nu = 1$ or 2 hold, then a $G_{n\nu}$ which is the intersection of a C_n with a Γ_ν is associated with respect to the curves $C_{n+\nu-3}$ of the plane, or rather it presents not more than $n\nu - 1$ conditions on a $C_{n+\nu-3}$ to contain it. Conversely, each $G_{n\nu}$ which is associated with respect to the $C_{n+\nu-3}$'s, and which lies on an irreducible C_n, is the complete intersection of the latter with a Γ_ν; the existence of such a C_n is a priori assured for $n \geqq 2\nu - 3$.

If, in order to fix our ideas, we suppose that $n \geqq \nu$, it follows that *the number c of distinct conditions to which a $G_{n\nu}$ must be subjected, in order that it should be the intersection of a C_n with a Γ_ν, is:*

$$c = (n-1)(n-2) \text{ if } n = \nu, \quad \text{and} \quad c = \nu(n-3) + 1 \text{ if } n > \nu .$$

Such conditions can be stated in various ways, and some of them will appear explicitly in the sequel. We add that, given any integer $l \geqq 1$, if in the foregoing we substitute for Γ_ν the curve $(l+1)\Gamma_\nu$ (i.e., Γ_ν counted $l+1$ times), the set $G_{n\nu}$ is replaced by the $n\nu$ differential elements E_j^l $(j = 1, 2, \ldots, n\nu)$, of order l, having as centres the various points of $G_{n\nu}$ and lying on C_n. These elements are then bound by certain conditions; and we see that *the number,* t_l, *of relations connecting the above* $n\nu$ *elements, mutually independent and independent of those connecting the* $n\nu$ *elements* E_j^{l-1}, *is given by* $t_l = l\nu^2 + \nu(\nu-3)/2$ (so that, for example, $t_l = l - 1$ if $\nu = 1$).

In order that the relations in question may be formulated explicitly, we introduce Cartesian coordinates (x, y) in the plane, referred to axes x, y which are generally situated with respect to the curve C and the set $G_{n\nu}$. We suppose, further, that $G_{n\nu}$ consists of $n\nu$ distinct points $P_j\,(x_j, y_j)$ in the finite part of the plane, and that C is irreducible and possesses n distinct points Q_i at infinity $(i = 1, 2, \ldots, n;\ j = 1, 2, \ldots, n\nu)$. Let

$$f(x, y) = 0 \quad \text{and} \quad \varphi(x, y) = 0 \tag{12}$$

be the equations of C and Γ respectively, and let $f^*(x, y)$ and $\varphi^*(x, y)$ be the forms of degree n and ν in $f(x, y)$ and $\varphi(x, y)$ respectively. Finally, let $\alpha(x, y)$ denote any polynomial with degree not greater than $n + \nu - 2$ in x, y, and let $\alpha^*(x, y)$ be the corresponding form of degree $n + \nu - 2$ (so that $\alpha^* = 0$ if α has degree $\leqq n + \nu - 3$). We consider the Abelian integral

$$\mathcal{J} = \int \frac{\alpha(x, y)\, dx}{f_y(x, y)\, \varphi(x, y)} \tag{13}$$

attached to C. It is immediately seen that the singular points of $\mathcal{J}$ can only occur at P_j and Q_i $(j = 1, 2, \ldots, n\nu;\ i = 1, 2, \ldots, n)$, and at these points its residues are respectively $\alpha(x_j, y_j)/J(x_j, y_j)$ and $-\alpha^*(1, t_i)/[f_y^*(1, t_j)\, \varphi^*(1, t_j)]$, where $J(x, y) = \partial(\varphi, f)/\partial(x, y)$, and t_i denotes the root of the equation $f^*(1, t) = 0$, which gives the angular coefficient of the improper point Q_i. Consequently, we have the identity

$$\sum_{j=1}^{n\nu} \frac{\alpha(x_j, y_j)}{J(x_j, y_j)} = \sum_{i=1}^{n} \frac{\alpha^*(1, t_i)}{f_y^*(1, t_i)\, \varphi^*(1, t_i)}\,; \tag{14}$$

and this reduces to the equation of Jacobi

$$\sum_{j=1}^{n\nu} \frac{\alpha(x_j, y_j)}{J(x_j, y_j)} = 0\,, \tag{15}$$

on the hypothesis that $\alpha(x, y)$ has degree $\leqq n + \nu - 3$.

Equation (15) furnishes a system of relations, since it contains the coefficients of $\alpha(x, y)$ as arbitrary parameters. It implies that a curve $\alpha(x, y) = 0$ with order $n + \nu - 3$, passing through $n\nu - 1$ points of $G_{n\nu}$, must contain $G_{n\nu}$ entirely; it therefore gives precisely the required conditions for $G_{n\nu}$. It will be seen that (15) still holds when C meets the line at infinity in points not all distinct. Moreover, if, by varying C or Γ, two or more points P_i become coincident at a finite point, P_0, then J becomes zero at that point; so that each of the corresponding terms of the summation in (15) tends to infinity. However, the sum of these terms has a determinate finite limit, which is obtainable as a residue of the integral (13) at the point P_0. This allows us in every case to replace (15) by another equation characterizing $G_{n\nu}$. Thus, for example, if C, Γ have simple contact at P_0, then the expression replacing the two terms in (15) which become infinite is the value at P_0 of

$$\frac{2\{\alpha, f\}}{\{J, f\}} - \frac{2\alpha}{3\{J, f\}^2} \times$$
$$\times \left[J_{xx} f_y^2 - 2 J_{xy} f_x f_y + J_{yy} f_x^2 + (\{\varphi_x, f\}\{f_y, \varphi\} - \{f_x, \varphi\}\{\varphi_y, f\}) f_y/\varphi_y\right],$$

where $\{u, v\}$ denotes the Jacobian determinant of the functions u, v with respect to x, y.

In an analogous manner we can work with the more general relation (14). If, for example, we assume that

$$\alpha(x, y) = a f_x \varphi_x + b f_x \varphi_y + c f_y \varphi_x + d f_y \varphi_y,$$

where a, b, c, d are four arbitrary constants, then the right-hand side of (14) (and consequently also the left-hand side) depends only on the points at infinity of C and Γ. Choosing, in particular, $a = b = 1$, $c = d = 0$, we obtain without difficulty the following proposition, due to HUMBERT [56], which contains as a part the final theorem of § 37, p. 80.

If two algebraic curves of an Euclidean plane meet in distinct finite points only, then the sum of the cotangents of the angles at which they cut at these points depends only on the points at infinity of the two curves. This sum is zero if one of the two curves is a circle, or, more generally, a curve of order $2m$ which cuts the line at infinity in the circular points each counted m times.

§ 39. The relation of Reiss, and some extensions. — We consider again the two curves C_n, Γ_ν, generally situated, represented by (12). From the previous section, in order to obtain all the *relations connecting the differential elements E_j^l* — defined by the curves — it suffices to proceed in the manner which will now be specified. Firstly, instead of (13), we consider the Abelian integral

$$\mathcal{J}_l = \int_C \frac{\alpha(x, y)\, dx}{f_y(x, y)\, \varphi^{l+1}(x, y)},$$

where $\alpha(x, y)$ now denotes any arbitrary polynomial with degree $\leq n + (l + 1)\nu - 3$ in x, y. We obtain the required relations by equating to zero the sum of the residues of $\mathfrak{J}_l$ for every choice of α, and by eliminating between the resulting relations the derivatives with orders greater than l of the functions f and φ, computed at the points P_j. The validity of this procedure is maintained, except for conceptually inessential modifications, however the curves C and Γ are specialized; moreover, it furnishes incidentally some of the relations connecting the differential elements with orders greater than l.

Hence, to give a simple example, if we assume that $l = 1$, $\alpha = (\varphi\varphi_{yy} - \varphi_y^2) f_y$, and that C, Γ are generally situated, then we obtain the identity

$$\sum_{j=1}^{n\nu} \frac{y_j''(x_j) - \eta_j''(x_j)}{[y_j'(x_j) - \eta_j'(x_j)]^3} = 0\,, \tag{16}$$

where $y = y_j(x)$ and $y = \eta_j(x)$ are respectively the local equations of C and Γ in the neighbourhood of the point $P_j(x_j, y_j)$.

Equation (16) could easily be generalized in several directions. If, as a very special case, we choose Γ to be a line $y =$ const., generally situated with respect to C, then (16) reduces to the relation of Reiss $\sum_{i=1}^{n} y_i''/y_i'^3 = 0$. This relation can also be given in the metrical form

$$\sum_{i=1}^{n} \frac{K_i}{\sin^3 \tau_i} = 0\,, \tag{17}$$

where τ_i denotes the angle at which C_n meets the line Γ at P_i, and K_i denotes the curvature of C_n at this point; this is the necessary and sufficient condition that n E_2's with collinear centres should belong to a C_n, which does not contain the line of centres as a component. We remark that, in conformity with what has been said in § 38, p. 81, we have a means of extending the range of (16) and (17), such that they can be used however C and Γ are mutually situated at their common points.

Thus, for example, assuming that Γ coincides with the x-axis, we obtain, in all circumstances, a relation (which reduces to (17) for the case in which C and Γ meet in n distinct points) by *equating to zero the sum of the residues of the Abelian integral* $\int_C \frac{dx}{y^2}$ *at the points of intersection of* C *and* Γ, *relative to the different branches of* C *with centres at those points;* such residues we shall refer to simply as the *residues of the corresponding branches.* By means of a calculation free from difficulty, we then deduce that *a branch with order* p *and class* q *makes no contribution, if* $p < q$, *when the branch does not touch* Γ, *since its residue is then zero; it has a residue* ϱ *depending on the neighbourhood of order* $2p$ *on the branch, if* $p \geqq q$ *and if the branch does not touch* Γ*; and the residue* ϱ *depends on the neighbourhood of order* $2p + 3q$, *if the branch touches* Γ.

In the two final cases, we consider any algebraic curve, $\bar{C}$, which contains the neighbourhood indicated and has order $p+1$ or $p+q+1$ respectively. $\bar{C}$ then meets Γ in one, and only one, point distinct from the centre of the branch; such a point will be the origin of an $\bar{E}_2$ of $\bar{C}$ with residue $-\varrho$. We thus obtain, for the various $\bar{C}$'s only ∞^2 $\bar{E}_2$'s, each of which we shall say is *associated* with the corresponding branch relative to the line Γ. Such $\bar{E}_2$'s can also be characterized by the following obvious property: *it is equivalent to saying that an E_2 with centre on Γ and not touching Γ has residue ϱ, or that E_2 lies on a conic with one, and therefore with any, of those $\bar{E}_2$'s.* In virtue of the theorem of REISS (p. 84) and of the above extension, we see that:

Given in a plane a line Γ and a curve C_n not containing Γ as a component, we consider the various branches of C_n with centres on Γ, excluding the branches not touching Γ with classes exceeding their orders. If m ($\leqq n$) is the number of such branches, then m second order elements (with centres on Γ) associated with those branches, always belong to some algebraic curve with order m, which does not contain Γ as a component.

We shall show finally — by means of a convenient procedure relying only on the operation of differentiation — how, *from every identity of the form*

$$\sum_{j=1}^{n\nu} \varrho(x_j, y_j) = 0\,, \tag{18}$$

where $\varrho(x_j, y_j)$ is a rational function of the coordinates x_j, y_j of the point P_j, the coefficients of which depend on the differential elements of order l on C and Γ with centres at the points P_j, we can deduce several similar identities with respect to the elements of order $l+1$.

To this end, we choose arbitrarily two polynomials $f_1(x, y)$, $\varphi_1(x, y)$ with degrees not greater than n, ν, of which at most one may possibly vanish identically; and we consider the curves

$$F(x,y) = f(x,y) + \lambda f_1(x,y) = 0, \quad \Phi(x,y) = \varphi(x,y) + \lambda\varphi_1(x,y) = 0\,.$$

When the absolute value of λ is sufficiently small, the curves meet in $n\nu$ points $(X_j(\lambda), Y_j(\lambda))$ near to the points P_j ($j = 1, 2, \ldots, n\nu$). Applying (18) to the present case, we obtain a relation of the form $\Sigma R(X_j, Y_j) = 0$, which is valid identically in a neighbourhood of $\lambda = 0$. We note that, for $\lambda = 0$, we clearly have $X_j = x_j$, $Y_j = y_j$ and, moreover, $F_x = f_x$, $F_y = f_y$, $F_{xx} = f_{xx}$, etc.; so that — again for $\lambda = 0$ — we have

$$\frac{\partial R(X_j, Y_j)}{\partial X_j} = \frac{\partial\varrho(x_j, y_j)}{\partial x_j}\,, \qquad \frac{\partial R(X_j, Y_j)}{\partial Y_j} = \frac{\partial\varrho(x_j, y_j)}{\partial y_j}\,.$$

If now we put $\lambda = 0$ in the equations which we obtain by differentiating with respect to λ the identities

$$f(X_j, Y_j) + \lambda f_1(X_j, Y_j) = 0,\ \varphi(X_i, Y_j) + \lambda\varphi_1(X_j, Y_j) = 0,\ \sum_{j=1}^{n\nu} R(X_j, Y_j) = 0,$$

and then eliminate the $2n\nu$ quantities $X_j'(0)$, $Y_j'(0)$ between the $2n\nu+1$ equations so obtained, we emerge with the relation

$$\sum_{j=1}^{n\nu}\left[\frac{\{\varrho,\varphi\}\,f_1-\{\varrho,f\}\,\varphi_1}{\{\varphi,f\}}\right]_{P_j}=0\,,$$

which is of the required type. We recall (pp. 85, 83) that f_1 and φ_1 are two arbitrary polynomials with degrees not greater than n, ν, and that the brackets signify Jacobian determinants.

In the particular case for which $\nu=1$ (and n is any integer greater than 1), we can assume that $f_1\equiv 0$ and $\varphi_1\equiv x, y, 1$. Then the foregoing procedure, applied to (18), furnishes the three relations

$$\sum_{j=1}^{n\nu}\left[\frac{\{\varrho,f\}\,x}{\{\varphi,f\}}\right]_{P_j}=0\,,\quad \sum_{j=1}^{n\nu}\left[\frac{\{\varrho,f\}\,y}{\{\varphi,f\}}\right]_{P_j}=0\,,\quad \sum_{j=1}^{n\nu}\left[\frac{\{\varrho,f\}}{\{\varphi,f\}}\right]_{P_j}=0\,.$$

However, these reduce to only two distinct relations, since we obtain the identity when we sum the relations, having first multiplied them by the numbers a, b, c which appear in the equation $ax+by+c=0$ of Γ. In an analogous manner we see that, if we apply the same procedure to the two relations just obtained, we derive only three new relations, and so on.

Hence, commencing with the relation of REISS connecting the second order elements E_j^2, we obtain $l-1$ *distinct relations* connecting the n elements E_j^l with order l ($\geqq 2$) (having collinear centres, and lying on a plane curve C_n) which are independent of the relations connecting the elements of order $l-1$ with the same centres. As a consequence of § 38, p. 82, *this process furnishes all the t_l relations connecting such elements.*

§ 40. Further algebro-differential properties. — We shall conclude the present Part by outlining a number of intersection problems which lead to algebro-differential properties, of a type rather different from those considered in the previous sections, but with which they have a certain affinity.

In the real projective plane, we shall now mean by a "curve" any finite number of circuits (a circuit being a topological image of a circle). An important problem is that of discovering what further conditions should be added to ensure the algebraicity of such a "curve", that is, in order to be able to assert that it coincides with, or lies on, an algebraic curve of a given order. There is, of course, a great latitude in the choice of these conditions, which suggest investigations in different directions. However, the choice should be such that it might possibly lead to results which are simple and not trivial. An example is given by the following theorem.

There exist in the plane "curves" which meet the lines and the conics in not more than three and not more than six points respectively, these maxima being in fact attained, and which also do not coincide with — or lie on — plane cubic curves. But a "curve" which meets the plane cubics, not containing it, in at most nine points, necessarily coincides with — or lies on — a real plane algebraic curve of order three.

It is worthy of note that no hypotheses of differentiability enter into the above enunciation, the only conditions being those regarding the number of certain intersections. However, it seems probable that, should we wish to obtain further results of this nature, it would be necessary to introduce simultaneously suitable conditions of differentiability as well as on the intersections. This is illustrated by the following example concerning, however, curves in space.

In a three-dimensional (real or complex) space, we consider six linear branches L_i ($i = 1, 2, \ldots, 6$) — of differential class C^1 — with centres lying in a plane α which touches none of the branches. Every plane π sufficiently close to α cuts L_i in one, and only one, point P_i; we then have:

In order that the six branches L_i should all lie on a quadric, which does not touch the plane α, it is necessary and sufficient that — for any choice of π in the neighbourhood of α — the six points P_i should lie on an irreducible conic.

From this it follows that:

The algebraic curves of order $\geqq 6$ lying on a quadric are the only skew algebraic curves which admit ∞^3 6-secant conics.

However, it does not seem possible to characterize in a similar manner the curves which lie on a surface of order greater than or equal to three. In fact, for example,

There exist skew algebraic curves of order 10 which lie on no cubic surface, but which cut the generic plane of space in ten points lying on a plane cubic.

The problem of characterizing the algebraic curves, by means of properties of an algebro-differential nature, is manifestly linked to that of individualizing such curves by associating suitable elements of this last type.

Thus, for example, we have the following theorem.

An algebraic curve of order $n \geqq 5$, which lies on an assigned quadric, and which meets the generators of one system in $r \geqq 1$ points and those of the other in $n - r \geqq r$ points, is determined uniquely when we prescribe on the quadric, in a general manner:

n (distinct) points of a plane section; the E_1's having $n - 1$ of these points as centres; the E_2's relative to $n - 3$ of the E_1's, etc.; finishing with $n - 2r + 1$ E_r's (containing $n - 2r + 1$ of the $n - 2r + 3$ E_{r-1}'s already fixed).

This result enables us to count the number of distinct relations connecting the differential elements of the above curve, which are chosen to contain a set of sub-elements of the type just described. Should we wish to determine effectively such relations, we could first reduce to the plane case by means of a stereographic projection of the quadric, and then apply — with suitable modifications — the methods of §§ 38, 39, pp. 81—86.

Historical Notes and Bibliography

The considerations and the developments of § 34, p. 69, concerning the residues of correspondences between curves, are already to be found in B. SEGRE [105] apart from the geometric interpretation of those residues, which appears here for the first time in such a general, explicit form. Some preliminaries on this topic are already in B. SEGRE [101] and [105]. The important complement of the correspondence principle of § 35, p. 74, is derived from B. SEGRE [104, 105]. From the same author also descend the geometric characterizations of the residues of the Abelian integrals which are here repeated in § 36, p. 76 (cf. B. SEGRE [111]), and again the applications made in § 37, p. 79 (cf. B. SEGRE [101]). With regard to these applications, it should be noted, however, that a particular case of the final theorem of § 37 (when $\nu = 0$) is already to be found — proved in a different manner — in G. HUMBERT ([56], theorems II and IV).

For an exposition of the important role played in algebraic geometry by the identity of JACOBI, and for the ample relevant bibliography, see B. SEGRE [113]. Some of the results of § 38, p. 81, are due to the latter author, as is the systematic derivation of the properties displayed there from the theory of residues. The same is true of the properties considered in § 39, p. 83, which all date from B. SEGRE [113] (together with further properties of the same type), excluding, of course, the simple case of the theorem of REISS. For the curious history of this theorem, and for the relevant bibliography, cf. BOMPIANI [33]. This author introduced the associated elements E_2, in the case of a rather simple type of branches [with the notation of § 39, p. 84, for precisely the cases $(p, q) = (1, 1), (1, 2), (2, 1)$], and he also extended the theorem of REISS to such branches.

The first theorem of § 40, p. 86, is taken from B. SEGRE [119], and the second from B. SEGRE [117]. The corollary to the second theorem, concerning the algebraic curves of a quadric, was reobtained by direct algebraic methods by E. MARCHIONNA [75], who first proved that an irreducible algebraic twisted curve — with order $n > 6$ — necessarily lies on a quadric, if each of its sections by the ∞^2 planes of a star (through the centre of which passes only a finite number of chords, none of these being trisecant for the curve) has six points lying on a conic. The last theorem of § 40, p. 87, is due to E. BOMPIANI [25] who has also, in a similar manner, studied the cases of the twisted quartics of the first and second species.

Part Six

Extensions to Algebraic Varieties

We shall now indicate the way in which many of the properties of curves, discussed in the preceding Part, can be extended to varieties of higher dimension. This is a vast field of research, which is by no means completely exploited, since several of the properties with which we shall deal could be extended further.

Moreover, the types of generalizations which we consider here are probably not the only extensions which are likely to prove fruitful.

§ 41. Generalizations of the equation of Jacobi. — Let us consider a system of $r \geqq 2$ algebraic equations in as many indeterminates:

$$f_1(x_1, x_2, \ldots, x_r) = 0,\ f_2(x_1, x_2, \ldots, x_r) = 0, \ldots, f_r(x_1, x_2, \ldots, x_r) = 0, \quad (1)$$

with degrees respectively $n_1, n_2, \ldots, n_r$, and let us suppose, at first, that the solution of the equations presents no speciality. More precisely, in an S_r, where $(x_1, x_2, \ldots, x_r)$ are interpreted as non-homogeneous coordinates, we suppose that the $r-1$ hypersurfaces $f_1 = 0, f_2 = 0, \ldots, f_{r-1} = 0$ meet in a curve, C, free from singular points and improper components, which cuts $f_r = 0$ in $N = n_1 n_2 \ldots n_r$ points, P_j, distinct and all finite, and which intersects the hyperplane at infinity in $N' = n_1 n_2 \ldots n_{r-1}$ distinct points, Q_i; we make the further, non-restrictive, hypothesis that the x_1-axis contains none of the points Q_i and is not parallel to the tangent to C at any of the points P_j. Putting, for brevity, $f_{hk} = \partial f_h / \partial x_k$, and denoting by J_{hk} the algebraic complement of f_{hk} in the Jacobian J of the f_h's with respect to the x_k's $(h, k = 1, 2, \ldots, r)$, we then have that at no point of C are $J_{r1}, J_{r2}, \ldots, J_{rr}$ zero simultaneously and that, at each point Q_i, $J_{r1} f_r \neq 0$. Moreover, along C we have the relations:

$$dx_1 : dx_2 : \ldots : dx_r = J_{r1} : J_{r2} : \ldots : J_{rr}.$$

Let α be any polynomial in the x's with degree not greater than $n_1 + n_2 + \cdots + n_r - r$, and consider the Abelian integral:

$$\mathcal{J} = \int\limits_C \frac{\alpha\, dx_1}{J_{r1} f_r} = \int\limits_C \frac{\alpha\, dx_2}{J_{r2} f_r} = \cdots = \int\limits_C \frac{\alpha\, dx_r}{J_{rr} f_r},$$

attached to C. It is seen immediately that the integral possesses no singular points other than those at the points P_j and Q_i; and so, using the fact that the sum of the residues is zero, we obtain the following identity, which generalizes the identity of JACOBI (§ 38, p. 82):

$$\sum_{j=1}^{N} \left(\frac{\alpha}{J}\right)_{P_j} = \sum_{i=1}^{N'} \left(\frac{x_1 \alpha}{J_{r1} f_r}\right)_{Q_i}. \quad (2)$$

The right-hand side of (2), with the present hypotheses, is finite, depends only on the varieties consisting of the improper points of the hypersurfaces (1), and is identically zero if the degree of α is not greater than $n_1 + n_2 + \cdots + n_r - (r+1)$. We note that in the remaining case, when α has degree precisely $n_1 + n_2 + \cdots + n_r - r$, the right-hand side formally depends on the order in which we have considered the hypersurfaces (1); however it does not, in fact, depend on this order, apart from the sign, since this is manifest for the left-hand side of (2). Thus we deduce from (2) a certain number of relations of another type, by equating the expressions given by the right-hand sides of (2) which correspond to the various ways of ordering (1).

We recall that (2) has been obtained on certain hypotheses of generality for (1); but most of these restrictions can be removed, by using a suitable continuity argument. Hence we can enunciate that:

If the hypersurfaces (1) *cut in a finite number of points* P_j, *which are all distinct and finite, and if* α *denotes a polynomial in the x's with the degree* $\leq \Sigma n_i - r$, *then* $\Sigma(\alpha/J)_{P_j}$ *depends only on the varieties consisting of improper points of the above hypersurfaces, and is zero if* α *has degree* $\leq \Sigma n_i - (r+1)$.

The relation (2) can, for example, be applied when we assume:

$$\alpha(x_1, x_2, \ldots, x_r) = \sum_{(h)} a_{h_1 h_2 \ldots h_r} f_{1h_1} f_{2h_2} \ldots f_{rh_r},$$

with the coefficients, a, arbitrary constants not all zero. Then the left-hand side of (2) — altered by a convenient constant factor $k \neq 0$ — admits an intrinsic geometric significance. We can prove, in fact, that $k(\alpha/J)_{P_j}$ is a *projective invariant of intersection of the hypersurfaces* (1) at the points P_j, taken together with a fixed *multilinear relation*, on the hyperplane at infinity in S_r, which is represented by placing the restriction $\Sigma a_{h_1 h_2 \ldots h_r} u_{1h_1} u_{2h_2} \ldots u_{rh_r} = 0$ on the r hyperplanes $u_{l1} x_1 + u_{l2} x_2 + \cdots + u_{lr} x_r = 0$ $(l = 1, 2, \ldots, r)$. From this, one could deduce several consequences of a projective or metrical nature.

The theorem just proved extends immediately to the case of *multiple intersections*, provided that these are all proper and finite in number. More precisely, if a point P_j absorbs $h > 2$ intersections of the varieties (1), the h corresponding terms of the sum in the left-hand side of (2) have to be substituted by the residue at P_j of the Abelian integral $\mathcal{J}$.

Further extensions can be obtained on the hypothesis that *the number of intersections becomes infinite*. Then various possibilities arise; but here we shall limit ourselves to a relatively simple case, which will suffice to suggest the power of the method employed and show how to proceed in other cases.

We suppose that *the varieties* (1) *intersect along a* V_d $(1 \leq d \leq r-1)$, free from multiple points, and also in a set of N $(< n_1 n_2 \ldots n_r)$ distinct points, P_j, none of which lies on V_d or at infinity. The $r-1$ hypersurfaces $f_1 = 0, f_2 = 0, \ldots, f_{r-1} = 0$ then cut, outside V_d, in a curve C: we suppose further that C is free from multiple points and that it meets the hyperplane at infinity of S_r in N' $(< n_1 n_2 \ldots n_{r-1})$ distinct points, Q_i. As before, we consider the Abelian integral $\mathcal{J}$, attached to C: in the present case, it can have no singular points other than the P_j's and Q_i's and the points, R, in which C meets V_d. At each such point R all the hypersurfaces of the linear system $\lambda_1 f_1 + \lambda_2 f_2 + \cdots + \lambda_{r-1} f_{r-1} = 0$ admit a certain S_{d+1} as tangent; so that R is a d-ple base-point for the linear system $\mu_1 J_{r1} + \mu_2 J_{r2} + \cdots + \mu_r J_{rr} = 0$. It follows that, if *the hypersurface* $\alpha = 0$ *passes through* V_d *with multiplicity* $\geqq d+1$, none of the points R is singular for $\mathcal{J}$, whence — with this hypothesis — *the validity of* (2) *is conserved* (except for new meanings for the symbols). This result can be easily extended further to the case in which C possesses some multiple points. Moreover, when concidences occur between the P_j's and Q_i's which appear in (2), this relation must be modified in a manner which is easily deduced from the foregoing.

§ 42. Generalizations of the relation of Reiss. — An analysis analogous to that of the previous section can be applied to suitable Abelian integrals, rather different in character from the integral $\mathcal{J}$ just considered, in order to obtain results generalizing in various ways those of § 39, p. 83. We shall here only give an extension of the relation of Reiss. We begin with the integral $\int_C d x_1 / f_r^2$, where the hypotheses and notation are the same as those at the beginning of § 41, p. 89. We put, for brevity:

$$J_i = \frac{\partial J}{\partial x_i}, \quad J_{hki} = \frac{\partial J_{hk}}{\partial x_i}$$

and, with the usual procedure, we obtain the identity:

$$\sum_{j=1}^{N} \left(\sum_{i=1}^{r} \frac{J_{r1} J_{ri} J_i - J J_{ri} J_{r1i}}{J^3} \right)_{P_j} = 0 , \tag{3}$$

which gives *a relation between the second order differential elements* (of dimension $r-1$) of the hypersurfaces (1) having as centres the N points of intersection P_j. We obtain other relations immediately from (3) by taking, as is allowed, in place of some of $f_1, f_2, \ldots, f_r$ — chosen so as to have the same degree — a general linear combination of them with constant coefficients.

In particular, we can apply (3) to the case in which $n_2 = n_3 = \cdots = n_r = 1$, so that $N = n_1$ and the P_j's are points of a line, l; it is then not restrictive to suppose that l coincides with the x_1-axis. Thus we can take

$$f_h = a_{h2} x_2 + \cdots + a_{hr} x_r \quad (h = 2, \ldots, r) ,$$

where the a_{hk}'s are $(r-1)^2$ constants with the determinant $A \neq 0$. We suppose, moreover, that f_1 is any polynomial of degree n ($\geqq 2$) in $x_1, \ldots, x_r$, having n distinct roots for x_1 when $x_2 = \cdots = x_r = 0$; this polynomial we shall also indicate by φ, and we shall denote its derivatives by the appropriate suffixes. The relation (3) now reduces to a homogeneous quadratic relation in the algebraic complements A_{rh} of the a_{rh}'s in the above determinant A; equating to zero the coefficient of $A_{rh} A_{rk}$, we obtain the identity:

$$\sum_{j=1}^{n} \left[\frac{\varphi_h \varphi_k \varphi_{11} - \varphi_1 (\varphi_h \varphi_{1k} + \varphi_k \varphi_{1h}) + \varphi_1^2 \varphi_{hk}}{\varphi_1^3} \right]_{P_j} = 0 \quad (h, k = 2, 3, \ldots, r) . \tag{4}$$

We now have altogether $r(r-1)/2$ *relations*, which we can prove to be not only *necessary*, but also in general *sufficient*, in order that n *second order (hypersurface) elements of* S_r — *with centres on a line* l *and not touching* l — *should lie on a hypersurface of order* n *not passing through* l.

We note that each of the expressions which appear in (4) between the square brackets is covariant under the translations along the x_1-axis. Taking account of the projective character of the complex of the above relations among the differential elements, we deduce that:

If n *second order elements with centres on a line* l *belong to a hypersurface of order* n *in* S_r, *not passing through* l, *then the same property is enjoyed by the elements derived from the given ones by submitting each of them to an arbitrary special homology with the centre on* l.

We now propose to give geometric meanings for the $r(r-1)/2$ conditions (4). A first geometric interpretation follows from that already obtained in § 39, p. 84 (for $n=2$), when we observe that (4) are equivalent to the equations which we obtain by applying the relation of Reiss to l and the curve cut by the hypersurface $\varphi=0$ on a variable plane containing l. Other, more direct, interpretations are obtained as follows; firstly for the least values of n, with, of course, the hypothesis that $r \geqq 3$.

For $n=2$, it is evident that two second order elements of S_r lying on a quadric, which does not contain the line joining them, must have their *asymptotic cones cutting in the same* V^2_{r-3} *the space* S_{r-2} *which is the polar of* l *with respect to the quadric* (such a V^2_{r-3} being the section of S_{r-2} with the quadric). This furnishes $(r-2)(r+1)/2$ distinct conditions on the two elements; so that there will be only one further condition, and this will permit — when the curvature of one of the two elements is known, for which the asymptotic cone is already fixed — the determination of the curvature of the other, relative to which the asymptotic cone is also consequently determined. This condition can be given in the metrical form

$$\frac{k_1}{\cos^{r+1}\alpha_1} = (-1)^{r+1}\frac{k_2}{\cos^{r+1}\alpha_2}, \tag{5}$$

where k_1, k_2 are the curvatures of the two elements, and α_1, α_2 denote the angles determined by l and the normals of the elements. In a projective form, the complex of $r(r-1)/2$ conditions in question is equivalent to the following. *The two elements must be interchanged by the harmonic homology which has, as centre, any point on* l *distinct from the centres of the elements, and which transforms the tangent prime of one into that of the other.*

For $n=3$, we consider a cubic hypersurface, φ, and three of its second order elements with centres O', O'', O''' lying on a generic line l.

φ and the cubic hypersurface which consists of the three hyperplanes π', π'', π''' touching the differential elements at their centres, determine a pencil which contains a hypersurface, ψ say, passing through l. For such a ψ, each of the three points O', O'', O''' is manifestly double, whence ψ admits the line l as locus of double points. The first polars of the single points of l with respect to ψ form a pencil, which consists of the quadric cones which are tangent to ψ at these points; further, it is clear that those corresponding to the points O', O'', O''' cut, respectively, on π', π'', π''' the asymptotic cones of the three elements. Therefore, in order that three second order elements with centres on a line l, and not touching l, should lie on a cubic hypersurface which does not pass through l, it is necessary that *the quadric cones projecting their asymptotic cones from l should belong to a pencil.* This furnishes in all $(r-2)(r+1)/2-1 = r(r-1)/2-2$ conditions for the three differential elements; the remaining two, determine the mutual ratios of curvatures.

For $n \geqq 4$, we can proceed inductively, using the cases already investigated, when we have observed that, in view of (4), the following theorem holds.

Given n second order elements, with centres on a line l which does not touch any of them, in order that there should exist a hypersurface of order n which contains them, it is necessary and sufficient that — dividing the n elements (in some manner and, consequently, in any manner) in m sets of $i_1, i_2, \ldots, i_m$ elements, where $2 \leqq m \leqq n-1$ and $i_1+i_2+\cdots+i_m = n$ — if we construct in a generic way m hypersurfaces of orders i_1+1, i_2+1, $\ldots$, i_m+1 which contains those groups respectively, then the m $(< n)$ residual elements of those hypersurfaces with centre on l belong to a hypersurface of order m not passing through l.

Finally, it is almost superfluous to note how the method of residues, which we have been using, could be made to deal similarly with the cases in which the hypersurface φ and the line l have arbitrary behaviour at their common points.

§ 43. The residue of an analytic transformation at a simple united point. — Let T be an ∞^n analytic correspondence of a complex V_n into itself, which admits a (simple) point O of V_n as a simple united point. This means that the variety representing T on the product $W_{2n} = V_n \times V_n$ is an analytic variety of dimension n, passing simply through $O \times O$ and not touching the diagonal of W there. It follows that — with any allowable coordinate system $(x_1, x_2, \ldots, x_n)$ on V_n in a neighbourhood of O, at which point we shall for convenience assume that the x's all vanish — in the neighbourhood of O the equations of T can be written in the form

$$f_h(x_1, x_2, \ldots, x_n; X_1, X_2, \ldots, X_n) = 0 \quad (h = 1, 2, \ldots, n), \tag{6}$$

where the X's are coordinates of the transform by means of T of a point x, and the f_h's are n analytic functions of the $2n$ indeterminates x, X, which vanish when the indeterminates are all zero; such that, moreover, putting

$$f'_{hk} = \frac{\partial f_h}{\partial x_k}, \quad f''_{hk} = \frac{\partial f_h}{\partial X_k}, \quad f_{hk} = f'_{hk} + f''_{hk} \qquad (h, k = 1, 2, \ldots, n), \tag{7}$$

$$D = \det [f_{hk}]_{h,k=1,2,\ldots,n}, \tag{8}$$

and denoting by a zero suffix the substitution of zero for each the x's and the X's, we have:

$$[D]_0 \neq 0. \tag{9}$$

With the hypotheses as above, and taking account of §§ 2, 3, 8, pp. 3, 5, 15 we can consider the residue of T at O: this is a complex number, which we shall denote by the letter ω (possibly with the same index or dashes as we attach to T). We propose to calculate ω when T is defined by means of (6).

Suppose first that

$$\det [f''_{hk}]_0 \neq 0. \tag{10}$$

We can then solve (6) with respect to the X's in the neighbourhood of O, so obtaining equations of the type (1) of § 8, p. 15. It suffices then to employ the expression for ω given at the end of § 8. Making explicit the derivatives of the functions X of the x's, defined by (6) (calculated at O), by means of the derivatives of the f_h's, we see that:

$$\omega = [(D_1 + D_2 + \cdots + D_n)/D]_0 = \left[\sum_{h=1}^{n} \sum_{k=1}^{n} f''_{hk} D_{hk}/D \right]_0, \tag{11}$$

where we have put

$$D_k = \det [f_{h1} \cdots f_{h,k-1}\, f''_{hk}\, f_{h,k+1} \cdots f_{hn}]_{h=1,2,\ldots,n}, \tag{12}$$

and where D_{hk} denotes the algebraic complement of the element f_{hk} in the determinant (8).

The right-hand side of (11) remains finite, even if the determinant (10) is zero. A simple continuity argument shows then that (11) holds even when (10) ceases to be valid, so that (11) *holds with the single hypothesis that* (9) *is valid.*

Let ω^* be the residue at O of the correspondence T^{-1}, the inverse of T; in view of (9), (11) we have

$$\omega^* = \left[\sum_{h=1}^{n} \sum_{k=1}^{n} f'_{hk} D_{hk}/D \right]_0,$$

so that

$$\omega + \omega^* = n. \tag{13}$$

Therefore:

If a united point of a correspondence and its inverse is simple, as a united point of one of them (and therefore also for the other), then the sum of their residues at that point is equal to the dimension of the variety on which they operate.

As a corollary, we have: *if, in the neighbourhood of O, the correspondence T has an involutory character, the residue of T at O is equal to $n/2$.*

§ 44. Some important particular cases. — In § 43, p. 94, we have established (11) under the single condition that relation (9) holds: this relation does not however exclude the fact that T might degenerate in some manner, in the neighbourhood of O. We shall now examine some characteristic cases, for each of which the hypotheses and notation of § 43, p. 93, will be conserved, unless notice is given explicitly to the contrary.

a) We suppose that, in the neighbourhood of O, the correspondence T *transforms the points of V_n into points of a fixed analytic $V_{n'}$*, passing simply through O $(1 \leqq n' \leqq n-1)$. There will then be, subordinate to T, in $V_{n'}$ an analytic correspondence, T', defined in an obvious way and called *the restriction of T to $V_{n'}$*: we shall see that T' is *an $\infty^{n'}$ correspondence having O as a simple united point*, and that *the residue ω' of T' at O is related to the residue of T by the simple equation*

$$\omega = \omega' + n - n'. \tag{14}$$

By suitable choice of the coordinates x in the neighbourhood of O on V_n, we can ensure that $V_{n'}$ has equations

$$x_{n'+1} = 0,\ x_{n'+2} = 0, \ldots, x_n = 0\,.$$

Then the equations (6) of T reduce to the form

$$f_h(x_1, x_2, \ldots, x_n; X_1, X_2, \ldots, X_n) = 0 \qquad (h = 1, 2, \ldots, n'),$$
$$X_{n'+1} = X_{n'+2} = \cdots = X_n = 0;$$

so that on $V_{n'}$ — where $x_1, x_2, \ldots, x_{n'}$ are admissible coordinates — the correspondence T' has the equations

$$f_h(x_1, x_2, \ldots, x_{n'}, 0, \ldots, 0; X_1, X_2, \ldots, X_{n'}, 0, \ldots, 0) = 0 \quad (h = 1, 2, \ldots, n').$$

Adopting for T' a notation (with dashes) analogous to that used for T, we obtain

$$D' = D,\ D_1' = D_1,\ D_2' = D_2,\ \ldots,\ D_{n'}' = D_{n'},\ D_{n'+1} = \cdots = D_n = D\,.$$

Therefore (9) implies a similar relation for T', which proves the first part of the assertion. Further, from (11) and a similar equation relative to T', we prove immediately the remainder of the assertion, given by (14).

b) Suppose that T *transforms the separate points of* V_n *into the point* O; then the equations of T can be written in the form $X_1 = X_2 = \cdots = X_n = 0$; so that, *at present,* $\omega = n$.

Applying property (a) or (b) to T^{-1}, and employing the result of § 43, p. 95, which is a consequence of (13), we obtain immediately the following two propositions.

c) Suppose that T — in the neighbourhood of O — *operates* (not on all the points of V_n, but) *only on the points of an analytic* $V_{n'}$ $(1 \leq n' \leq n-1)$ passing simply through O. Then *the restriction of* T *to* $V_{n'}$ *has at* O *the same residue as* T.

d) A transformation T, *totally degenerate,* in the sense that it transforms O into any variable point of V_n, has at O a *simple united point and zero residue.*

e) The results contained in b) and d) are easily extended *to degenerate correspondences of species* m, which are correspondences T associating with each point of a first variety V_m, fixed in V_n, an arbitrarily variable point of a second variety V_{n-m} of V_n $(0 \leq m \leq n)$. We deduce first, without difficulty, that T is analytic if, and only if, V_m, V_{n-m} are also analytic; that the united points of T are precisely the common points of V_m, V_{n-m}; finally that any one of these points, O, is a simple united point of T if, and only if, it is a simple intersection of V_m, V_{n-m}. With all these hypotheses, the residue of T at O is equal to that of the restriction of T to V_m, in view of c); but this residue is equal to precisely m, in virtue of b), so that:

Any degenerate correspondence of species m *admits the residue* m *at any of its simple united points.*

f) Suppose now that V_n is immersed in any analytic variety W_r, with dimension $r > n$, in which an analytic correspondence, K, of dimension $2r - n$ is given. *The restriction* T *of* K *to* V_n is an analytic correspondence, generally ∞^n, obtainable also as the *restriction to* V_n *of the* r*-dimensional correspondences* K' *and* K'', defined in the following manner on W_r. We consider the degenerate ∞^{r+n} correspondences (mutually inverse) which transform respectively any point of V_n into any point of W_r, and any point of W_r into any point of V_n; the intersections of K with these are respectively the K', K'' above.

We see immediately that a simple point O of V_n is a united point of T if, and only if, it is also united for K', K''; and that a united point O, simple for one of T, K', K'', is also simple for the other two. On this hypothesis, putting respectively ω, ω', ω'' for the residues of T, K', K'' at O, *we have*

$$\omega = \omega', \qquad \omega = \omega'' + n - r. \tag{15}$$

This follows immediately, in fact, from c), a) applied to T, when this is considered as the restriction of K' and K'' to V_n.

The formal content of (15) is clear from (11), but it is worth while to make it explicit. Taking local coordinates $(x_1, x_2, \ldots, x_r)$ on W_r, we suppose that

$$\varphi_i(x_1, x_2, \ldots, x_r) = 0 \qquad (i = 1, 2, \ldots, n-r) \quad (16)$$

are the local equations of V_n. If T is now represented on V_n by equations of the form

$$f_h(x_1, x_2, \ldots, x_r; X_1, X_2, \ldots, X_r) = 0 \qquad (h = 1, 2, \ldots, n), \quad (17)$$

the united point O is simple if (9) holds, where now we put

$$\varphi_{ij} = \frac{\partial \varphi_i}{\partial x_j} \qquad (j = 1, 2, \ldots, r)$$

and

$$D = \det \begin{bmatrix} f_{hj} \\ \varphi_{ij} \end{bmatrix}.$$

The first relation of (15) expresses the residue ω of T at O by means of the equation

$$\omega = [(D_1 + D_2 + \cdots + D_r)/D]_0,$$

in which

$$D_j = \det \begin{bmatrix} f_{h1} \cdots f_{h,j-1} & f''_{hj} & f_{h,j+1} \cdots f_{hr} \\ \varphi_{i1} \cdots \varphi_{i,j-1} & 0 & \varphi_{i,j+1} \cdots \varphi_{ir} \end{bmatrix}.$$

§ 45. Relations between residues at the same point. — We return to the hypotheses and notations of § 43, p. 93. Let us take any combination $i_1, i_2, \ldots, i_s$ of the numbers $1, 2, \ldots, n$, where s denotes an integer satisfying $1 \leqq s < n$. We divide equations (6) into two partial systems, one given by (6*), i. e. the equations we obtain from (6) for $h = i_1, i_2, \ldots, i_s$, and the other given by the remaining $n-s$ equations (6). In the second set we put $X_1 = x_1, X_2 = x_2, \ldots, X_n = x_n$, and so obtain a system representing on V_n an *analytic variety, of dimension s*, with a simple point at O; this variety we shall denote by $V_s^{(i_1 i_2 \cdots i_s)}$. On this variety, equations (6*) represent an ∞^s *correspondence*, having a simple united point at O; we denote this correspondence by $T_{i_1 i_2 \cdots i_s}$, and its residue at O by $\omega_{i_1 i_2 \cdots i_s}$.

We first examine the case in which $s = 1$. Then we obtain in the above manner, on V_n, n curves $V_1^{(h)}$ $(h = 1, 2, \ldots, n)$ passing through O in mutually independent directions; and, on each curve $V_1^{(h)}$, we obtain an analytic correspondence, T_h, with O a simple united point and with residue ω_h at O. We shall prove that *the following equality holds*:

$$\omega = \omega_1 + \omega_2 + \cdots + \omega_n. \quad (18)$$

It in fact suffices to apply § 44, p. 96, f) to T_h, in order to see that

$$\omega_h = [(D_1^h + D_2^h + \cdots + D_n^h)/D]_0,$$

where D is given by (8) and D_k^h is equal to precisely $f''_{hk} \cdot D_{hk}$; (18) now follows from (11).

We then apply the same procedure to the correspondence $T_{i_1 i_2 \cdots i_s}$, and we obtain on $V_s^{(i_1 i_2 \cdots i_s)}$ s curves, issuing from O, which *in fact coincide with the* $V_1^{(i_1)}, V_1^{(i_2)}, \ldots, V_1^{(i_s)}$ defined above; and, on these curves, we have likewise s correspondences, defined by $T_{i_1 i_2 \cdots i_s}$, which *are none other than* $T_{i_1}, T_{i_2}, \ldots, T_{i_s}$. From (18) then follows

$$\omega_{i_1 i_2 \cdots i_s} = \omega_{i_1} + \omega_{i_2} + \cdots + \omega_{i_s},$$

and therefore

$$\Sigma\, \omega_{i_1 i_2 \cdots i_s} = \binom{n-1}{s-1} \omega,$$

where the summation extends over all the combinations $i_1, i_2, \ldots, i_s$ of s numbers from $1, 2, \ldots, n$.

We arrive at the same results if we represent T (as at the end of § 44, p. 97) by (16) and (17), and if we define $V_s^{(i_1 i_2 \cdots i_s)}$, $T_{i_1 i_2 \cdots i_s}$ by combining the equations (16) of V_n with the equations obtained from (17) by a process similar to the one previously applied to (6). Indeed, for that purpose, it suffices to substitute for (17) the equations deduced from them by expressing the x's, X's as functions of n parameters which uniformize the neighbourhood of O on V_n, and then to recall the invariance property of the residues (§§ 2, 8, pp. 4, 15). Formula (18) has been recently generalized by ATIYAH and BOTT in their work on elliptic differential operators [3], [6], [7]; cf. [4], [5], [8]–[11].

§ 46. The total residues of correspondences of valency zero on algebraic varieties. — Let V_n be any irreducible algebraic variety free from multiple points; and let T denote an ∞^n algebraic correspondence of V_n with itself, possessing only isolated simple united points, which are therefore finite in number. We suppose that there are N such united points, $O_1, O_2, \ldots, O_N$. We shall also suppose that $n \geq 2$, since the case $n = 1$ has been already studied (§§ 34—37, pp. 69—80). The sum

$$\Omega(T) = \omega(O_1) + \omega(O_2) + \cdots + \omega(O_N)$$

of the residues of T at its united points will be called the *total residue* of T; it is *a priori* a complex number, invariant over the birational transformations of V_n which are free from exceptions, or, more generally, which are regular at the points $O_1, O_2, \ldots, O_N$. We note further that $\Omega(T)$ manifestly *depends continuously on* T. Putting $\Omega(-T) = -\Omega(T)$, we obtain, from the definition, the additive character of the total residues:

$$\Omega(T_1 \pm T_2) = \Omega(T_1) \pm \Omega(T_2);$$

moreover, recalling the final proposition of § 43, p. 95, we deduce

$$\Omega(T) + \Omega(T^{-1}) = n\,N. \tag{19}$$

We shall now prove the following:

Theorem 1. — *Each algebraic correspondence of valency zero (in the sense of* SEVERI [141, 142]*), having only simple united points, admits as total residue a whole number.*

The property is verified immediately for degenerate algebraic correspondences of any species, from § 44, p. 96, e). We shall prove it now for the generalized Zeuthen correspondences of valency zero (on this notion, cf. SEVERI [142] p. 109).

A transformation of this type can be defined by first considering V_n as a variety immersed in an S_r, and represented — in non-homogeneous coordinates $x_1, x_2, \ldots, x_r$ — by means of an algebraic system of equations, reducible in the neighbourhood of any fixed point on V_n to the form (16); thus T is represented on V_n by a system of algebraic equations of the type (17). For x fixed generally on V_n, equations (17) represent an M_{r-n} of S_r, cutting V_n in a finite number of points — the transforms of x by T — and further possibly in a fixed subvariety, A, (pure or impure) of V_n. There might be exceptional points x, for which it occurs that M_{r-n} cuts V_n outside A in an infinite number of points, but we shall suppose at first that in fact there exist no such points. Further, since T admits $O_1, O_2, \ldots, O_N$ as its united points, finite in number, we can add the hypothesis (which is not restrictive, in so far as it can always be satisfied by suitably modifying the representation of T) that none of these points lies on A.

Consequently, if we choose any number h from $1, 2, \ldots, n$, then the $n-1$ hypersurfaces

$$f_1(x;x) = 0, \;\ldots,\; f_{h-1}(x;x) = 0,\; f_{h+1}(x;x) = 0,\; \ldots,\; f_n(x;x) = 0$$

cut V_n, outside A, in a curve $V_1^{(h)}$; on this curve, the equation $f_h(x_1, x_2, \ldots, x_r; X_1, X_2, \ldots, X_r) = 0$ defines an algebraic correspondence of valency zero, T_h, having the above point O as simple united point. In virtue of § 35, p. 74, the total residue of this correspondence:

$$\Omega(T_h) = \omega_h(O_1) + \omega_h(O_2) + \cdots + \omega_h(O_N)$$

is an integer (equal to the 2-nd index of the correspondence T_h). We have also, in view of (18) of § 45, p. 97,

$$\Omega(T) = \sum_{k=1}^{N} \omega(O_k) = \sum_{k=1}^{N} \sum_{h=1}^{n} \omega_h(O_k) = \sum_{h=1}^{n} \Omega(T_h)\,, \tag{20_0}$$

whence follows immediately the assertion for the present case.

If T is now any correspondence of Zeuthen, by adding to T a suitable degenerate correspondence we obtain a correspondence, Θ_0, which is the complete limit of a variable Θ of the type just examined (cf. SEVERI [140], n. 4). By the continuity property already noted for total residues, we see that $\Omega(\Theta_0)$ is equal to $\Omega(\Theta)$, this being an integer (as we have proved immediately above) which therefore cannot change by varying Θ continuously. On the other hand, $\Omega(\Theta_0)$ differs from $\Omega(T)$ by the sum of the total residues of the above degenerate correspondences, each of which — as we know — is an integer. Hence $\Omega(T)$ is also an integer.

From the case of the Zeuthen correspondences just dealt with, we pass immediately to that of arbitrary correspondences of valency zero, by using the previously known fact that any such correspondence, T, can be expressed as the algebraic sum of a finite number of Zeuthen correspondences, and by remarking that each of the elements of such a sum can be assumed to possess only simple united points, when this happens for T.

We shall now make Theor. I more precise, by proving:

Theorem II. — *If a correspondence T of V_n with itself has valency zero and possesses only simple united points, then*

$$\Omega(T) = n\,\beta + \delta_1 + 2\,\delta_2 + \cdots + (n-1)\,\delta_{n-1}, \tag{20}$$

where (α, β) *are the indices and* $\delta_1, \delta_2, \ldots, \delta_{n-1}$ *are the ranks of* T.

In view of Theor. I, in order to calculate the total residue we can substitute for T any equivalent correspondence (on the product $V_n \times V_n$). It is known (Severi [129, 130]) that

$$T \equiv S_0 + S_1 + S_2 + \cdots + S_n\,, \tag{21}$$

where S_m is an algebraic sum of degenerate correspondence of species m, which may be assumed to have only simple united points ($m = 0, 1, \ldots, n$); so that the algebraic number N_m of united points of S_m is precisely equal to α for $m = 0$, to β for $m = n$, and to the rank δ_m for the intermediate values of m. From (21) we deduce that

$$\Omega(T) = \Omega(S_0) + \Omega(S_1) + \cdots + \Omega(S_m)\,,$$

and from § 44, p. 96 e), that $\Omega(S_m) = m\,N_m$; hence (20) follows.

We note finally that, since T^{-1} has indices (β, α) and ranks δ_{n-1}, $\delta_{n-2}, \ldots, \delta_1$, from (20) we obtain

$$\Omega(T^{-1}) = n\,\alpha + \delta_{n-1} + 2\,\delta_{n-2} + \cdots + (n-1)\,\delta_1\,.$$

And it suffices to add this, term by term, to (20), in order to obtain, with the use of (19), the *coincidence principle* of Severi [141, 142]:

$$N = \alpha + \beta + \delta_1 + \delta_2 + \cdots + \delta_{n-1}\,. \tag{22}$$

§ 47. The residues at isolated united points with arbitrary multiplicities. — We return to the consideration of an ∞^n analytic transformation T, of a complex V_n into itself, for which O is an isolated united point. To this point is attached, in a known manner, a positive integer k, the multiplicity of O as united point; the integer is unity if O is a simple united point, and, in general, it can be characterized as follows. If, as is certainly possible, T is made to vary a little with continuity in any manner, so that an ∞^n analytic correspondence T' of V_n with itself is obtained which in the neighbourhood of O possesses only simple united points (tending to O as T' tends to T), then the number of such united points is precisely k; we shall denote these points by $O_1', O_2', \ldots, O_k'$. We now define *the residue* $\omega(O)$ *of* T *at* O, by putting

$$\omega(O) = \lim_{T' \to T} [\omega'(O_1') + \omega'(O_2') + \cdots + \omega'(O_k')], \tag{23}$$

where $\omega'(O_i')$ denotes the residue of T' at O_i'. This definition is allowable, when we have proved that:

1. the limit stated on the right-hand side of (23) exists and is finite;
2. the limit is independent of the way in which T is varied to T'.

Further, it is preferable that

3. it should be possible to find a formula which permits, for a given T, the explicit calculation of $\omega(O)$.

As regards 1. and 2., we note that each of them involves only certain differential neighbourhoods of V_n and T at O, with sufficiently high orders. We can therefore (by substituting for V_n and T a variety and transformation which approximate to them suitably) suppose that V_n is an algebraic variety, and that T is an algebraic correspondence of V_n with itself, of valency zero, which possesses a finite number, N, of united points, of which k fall at O, whilst the remainder $O_{k+1}, \ldots, O_N$ are distinct, and therefore simple united points of T. Similarly, it is then not restrictive to suppose that T' is also algebraic of valency zero, and possesses N distinct united points. Hence, in virtue of § 46, p. 99, the total residue of T' is an integer, which does not depend on the way in which T is varied to T', and which can therefore be denoted by $\Omega(T)$. This, together with § 46, p. 98, and (23), furnishes easily

$$\omega(O) = \Omega(T) - [\omega(O_{k+1}) + \cdots + \omega(O_N)];$$

and this immediately establishes 1., 2.

We shall deal with 3., and again with 1., 2., on the hypothesis that it is possible to *represent* T *completely in the neighbourhood of* O *by analytic equations of type* (6); this hypothesis is rather general, but not fully so, in as much as there exist, manifestly, cases in which such a local representation is not possible.

We shall first examine the simple case $k = 1$, for which the above hypothesis is always satisfied (§ 43, p. 93); we know already that the residue $\omega = \omega(O)$ can then be calculated by means of (11). This equation can be written in a rather different form, when we have introduced the following analytic function (which is meromorphic in the neighbourhood of O):

$$\theta(x_1, x_2, \ldots, x_n) = \left[\frac{D_1 + D_2 + \cdots + D_n}{f_1 f_2 \cdots f_n}\right]_{X_1 = x_1, X_2 = x_2, \ldots, X_n = x_n}, \tag{24}$$

and when we have remarked that, from (7), (8), the denominator $D(0, \ldots, 0; 0, \ldots, 0)$ of (11) is simply the functional determinant of $f(x_1, \ldots, x_n; x_1 \ldots, x_n)$ with respect to the x's, calculated at the point O. Using results of B. SEGRE [97, n. 5], we deduce, without difficulty, that $\omega(O)$ *is equal to the residue at the point O of the analytic function* (24); thus it can be expressed by the equation

$$\omega(O) = \frac{1}{(2\pi i)^n} \int\int \cdots \int_{C_n} \theta(x_1, x_2, \ldots, x_n)\, d x_1\, d x_2 \ldots d x_n, \tag{25}$$

where the n-ple integral must be extended over an n-cycle, C_n, which lies in a neighbourhood of O on V_n, and which meets none of the hypersurfaces

$$f_h(x_1, x_2, \ldots, x_n; x_1, x_2, \ldots, x_n) = 0 \qquad (h = 1, 2, \ldots, n), \tag{26}$$

but has with this complex of hypersurfaces a suitable linking.

We now examine the case of $k \geqq 2$, and we begin with the initial definition of $\omega(O)$, that is, the definition by means of (23). In this case, the variation from T to T' is given by replacing the equations (6) of T by analogous equations

$$f'_h(x_1, x_2, \ldots, x_n; X_1, X_2, \ldots, X_n) = 0 \quad (h = 1, 2, \ldots, n), \tag{6'}$$

containing one or more parameters (not indicated explicitly), which tend to (6) when T' tends to T. We can now define the function $\theta'(x_1, x_2, \ldots, x_n)$ in the same way as we deduced (24) from (6). Thence the expression in square brackets in (23) is just the sum of the residues of $\theta'(x_1, x_2, \ldots, x_n)$ at the united points $O'_1, O'_2, \ldots, O'_k$, of T', which tend to O; thus it can be written in the form

$$\frac{1}{(2\pi i)^n} \int\int \cdots \int_{C'_n} \theta'(x_1, x_2, \ldots, x_n)\, d x_1\, d x_2 \ldots d x_n. \tag{27}$$

The n-ple integral is here extended over an n-cycle, C'_n, which is the sum of k n-cycles belonging respectively to the neighbourhoods of the points $O'_1, O'_2, \ldots, O'_k$ (and therefore lies similarly in the neighbourhood of O) on V_n, and having suitable linking at these points with the n hypersurfaces

$$f'_h(x_1, x_2, \ldots, x_n; x_1, x_2, \ldots, x_n) = 0 \qquad (h = 1, 2, \ldots, n). \tag{28}$$

Since, in a neighbourhood of O, the function θ' is holomorphic outside the hypersurfaces (28), the integral (27) is unchanged when C'_n varies homotopically without crossing these hypersurfaces. Whence — if T' is sufficiently near to T — we can substitute for C'_n in (27) a convenient homotopic transform, C_n, which belongs to the neighbourhood of O, but does not meet the hypersurfaces (26), and which remains unchanged when we vary T'. Such an n-cycle C_n will thus have with the hypersurfaces (26) a certain suitable linking, which it is not necessary to specify.

If now we pass to the limit, making T' tend to T, the expression (27) tends to the right-hand side of (25), so that (23) will be identifiable with (25). Hence we have *extended the validity of* (25) to the present more general case. This new formula (25) further gives a reply to 3., and also demonstrates immediately the truth of 1. and 2.

It should be noted that, in order to exhaust the study of 3., it would be necessary to assign explicitly the behaviour of the cycle C_n, which appears in (25), relative to the hypersurfaces (26). This would require a discussion of the various possible cases, and a complex topological analysis; but here we shall not study this any more deeply. We add only that a similar analysis might perhaps lead to an autonomous proof of (20), from which — as a consequence of § 46 — would follow a *topologico-functional proof* of the general principle of coincidence (22) [this would generalize the topological proofs of the Bézout theorem given recently by R. Caccioppoli (Ann. di Mat., (4) **29,** 1—14 (1949)), and E. Martinelli (Rend. di Mat., (5) **14,** 422—430 (1955))].

Finally we remark that, as a result of the definition of $\omega(O)$ given above for $k > 1$, it is meaningful to consider the *total residue* of any algebraic ∞^n correspondence T of V_n with itself, on the simple hypothesis that T should possess a *finite number of united points* (with arbitrary multiplicities). Then it is immediately seen that:

Theorems I and II of § 46, *pp.* 99, 100, *are still valid for every correspondence of valency zero of the above type.*

§ 48. Extensions to algebraic correspondences of arbitrary valency. — The results of § 46 can also be extended, as we shall now show, to all algebraic correspondences with valency.

The proof of Theor. I (of § 46) can be transferred, with obvious modifications, to the present more general case. But we can also proceed in the following way, when we wish to establish only that *for each T with valency, the total residue $\Omega(T)$ is either an integer or differs from an integer by* 1/2. It suffices to observe that, if T has valency, then the correspondence $T - T^{-1}$ has valency zero. But — from § 46, Theor. I —

$$\Omega(T - T^{-1}) = \Omega(T) - \Omega(T^{-1})$$

must be an integer. Hence, combining the result with (19), we deduce the assertion.

From this property, we see that, if *a correspondence T with valency* varies with continuity, having always a finite number of united points (which, taking account of the results of § 47, p. 100, we may suppose to have arbitrary multiplicities as united points), then *its total residue remains constant.*

This permits us *to assign intrinsically a value to $\Omega(T)$ for each correspondence T with valency*, even in the case for which T is virtual or has an infinity of united points. In fact, in each of the latter cases we can find an equivalence of the type

$$T \equiv T_1 - T_2,$$

where T_1 and T_2 are two correspondences with valency, both effective and possessing a finite number of united points. It is then allowable to assume as definition

$$\Omega(T) = \Omega(T_1) - \Omega(T_2),$$

since it is easily proved that the value for $\Omega(T)$ so obtained depends only on T, and not on the choice of T_1, T_2.

We then succeed in *extending the validity of* (19) *to the transformations of this type*, when the integer N which appears in (19) is regarded, of course, as the virtual number of united points of T. Thus, for example, considering the *identity correspondence*, S, of V_n into itself, we have $S^{-1} = S$; moreover, as is well known,

$$N = (-1)^n I_n + n - 1,$$

where I_n denotes, as usual, the ZEUTHEN-SEGRE invariant of V_n. Therefore the application of (19) to S furnishes

$$\Omega(S) = n\,[(-1)^n I_n + n - 1]/2. \tag{29}$$

We note that, from (29), *$\Omega(S)$ is always an integer*, since I_n is known to be even when n is odd (cf. B. SEGRE [94], n. 1).

Theor. II of § 46, p. 100, can now be extended to the case of a transformation T with arbitrary valency, γ, by substituting for it, as in (21), the analogous decomposition

$$T \equiv S_0 + S_1 + S_2 + \cdots + S_n - \gamma S,$$

with the obvious meaning for the symbols. Proceeding as in § 46, p. 100, and employing (29), we thus obtain:

The total residue of any correspondence T with valency, of an algebraic V_n with itself, is given by

$$\begin{aligned}\Omega(T) = n\,\beta + \delta_1 + 2\,\delta_2 + \cdots + (n-1)\,\delta_{n-1} \\ - n\gamma[(-1)^n I_n + n - 1]/2,\end{aligned} \tag{30}$$

where β, γ and $\delta_1, \delta_2, \ldots, \delta_{n-1}$ denote respectively the second index, the valency and the ranks of T, and I_n is the ZEUTHEN-SEGRE *invariant of V_n.*

We deduce from (30), with the aid of what has been remarked about (29), that $\Omega(T)$ *is always an integer*; and this extends the validity of Theor. I of § 46, p. 99, on the present more general hypotheses. Moreover, from (30), (19), we obtain immediately the following extension of (22) (which is also due to SEVERI [141, 142]):

$$N = \alpha + \beta + \delta_1 + \delta_2 + \cdots + \delta_{n-1} - \gamma\,[(-1)^n I_n + n - 1], \qquad (31)$$

for which one might possibly give an autonomous topological proof — of a different type — following the lines suggested in the antepenultimate paragraph of § 47, p. 103.

We note finally that, in view of the preceding remarks, (30) has a meaning and holds also for the case in which T is virtual, or possesses an infinity of united points. If T is an effective correspondence of this last type, then each irreducible component, M, of the algebraic variety consisting of the united points of T gives a contribution, $\omega(M)$, to the number $\Omega(T)$; such a contribution is a complex number, which is determined by a process analogous to that indicated in § 47, p. 103, for the proof of (25). The explicit expression for $\omega(M)$, of the type (25), would be most important, because — from (19) — the sum of $\omega(M)$ with the analogous contribution $\omega^*(M)$ relative to T^{-1} would give the multiple by n of the numerative equivalence of M in the number of united points of T given by (31).

§ 49. Applications to algebraic correspondences of a projective space into itself. — The results of §§ 46—48, pp. 98—105, can be applied and interpreted in divers ways, according to the particularity of the varieties and correspondences involved. We shall here single out the case of algebraic correspondences in *projective spaces* for further study; this leads to a marked simplification, since *each such algebraic correspondence has valency zero* (cf. SEVERI [139, n. 16]).

Let T be any algebraic correspondence of S_n into itself; T is then representable by a system of algebraic equations of the type (6). We denote by a_h, b_h the degrees of f_h in the x's, X's respectively $(h = 1, 2, \ldots, n)$. For simplicity, we shall consider only the case in which the system (6) is generic, in the sense that the properties specified in the next paragraph should hold for T.

The system (6), when we choose the set of the x's or of the X's generically, admits a finite number of solutions in the remaining variables; and none of these solutions is fixed, when we vary the generic set. This is equivalent to saying that the indices (α, β) of T can be expressed by

$$\alpha = a_1 a_2 \ldots a_n, \qquad \beta = b_1 b_2 \ldots b_n. \qquad (32)$$

The united points of T are then given by the equations

$$f_h(x_1, x_2, \ldots, x_n; x_1, x_2, \ldots, x_n) = 0 \quad (h = 1, 2, \ldots, n). \qquad (33)$$

This equation has degree $a_h + b_h$, and the set of the n equations admits a finite number of common solutions, distinct and all finite. The united points, O_k, of T are thus all simple and finite; in number, they are

$$N = (a_1 + b_1)(a_2 + b_2) \dots (a_n + b_n) . \tag{34}$$

In the present case, the rank δ_l of species l of T ($l = 1, 2, \dots, n-1$) is simply the order of V_l, the transform by T of the generic S_l of S_n (see SEVERI [142, n. 5]). It is therefore given by

$$\delta_l = \sum_{(i,j)} a_{i_1} a_{i_2} \dots a_{i_{n-l}} b_{j_1} b_{j_2} \dots b_{j_l} \quad (l = 1, 2, \dots, n-1), \tag{35}$$

where $(i_1, i_2, \dots, i_{n-l})$ and $(j_1, j_2, \dots, j_l)$ denote two complementary combinations of $n-l$ and l integers from $1, 2, \dots, n$, and the sum is taken over all such pairs (cf. VAHLEN [164]). As a check, (22) can be verified immediately from (32), (34) and (35).

From (6), by the methods of § 45, p. 97, we can define the algebraic curve $V_1^{(h)}$ and the algebraic correspondence T_h (for $h = 1, 2, \dots, n$); and we see immediately that, in the present case, T_h is an algebraic correspondence of $V_1^{(h)}$ into itself, having indices $(a_h N/(a_h + b_h), b_h N/(a_h + b_h))$ and valency zero. The united points of T_h are all simple, and coincide with the N points O_k. In view of § 35, p. 74, the total residue of T_h is accordingly given by

$$\Omega(T_h) = \sum_{k=1}^{N} \omega_h(O_k) = b_h N/(a_h + b_h) \quad (h = 1, 2, \dots, n). \tag{36}$$

Therefore, from (20_0) (p. 99), we have

$$\Omega(T) = \sum_{h=1}^{n} b_h N/(a_h + b_h) ; \tag{37}$$

whence (20) can be immediately verified, again with use of (32), (34) and (35).

Equations (36), and therefore also (37), assume the form of important *algebraic identities* (of the same type as that of JACOBI, obtained in § 41, p. 89), when the $\omega_h(O_k)$'s are expressed in the way indicated in § 44, p. 96, f). We then easily deduce the identity:

$$\sum_{k=1}^{N} \left[\frac{\sum_{j=1}^{N} f''_{hj} D_{hj}}{D} \right]_{O_k} = b_h N/(a_h + b_h) \qquad (h = 1, 2, \dots, n), \tag{38}$$

where the suffix O_k signifies that the expression between square brackets is to be evaluated by giving the x's and X's their values at O_k.

We now make the hypotheses that

$$a_1 = a_2 = \dots = a_n = a, \quad b_1 = b_2 = \dots = b_n = b. \tag{39}$$

If h and i denote two distinct fixed integers of the set $1, 2, \dots, n$, then we consider the correspondence, T_λ, defined by the system, (6_λ), deduced from (6) by replacing the equation $f_h = 0$ with $f_h + \lambda f_i = 0$. It is evident, from (39), that T_λ is of the same type as T (it is in fact identical with it when $\lambda = 0$), and that the two transformations possess the same united points; moreover, the determinant D_λ relative to (6_λ) does not depend on λ, as in fact it reduces to D. If we now write for T_λ the equation, (38_λ), analogous to (38), we see that the right-hand side is independent of λ, whilst the left-hand side contains λ linearly. Hence the coefficient of λ in this expression must be zero, so that we derive the identity

$$\sum_{k=1}^{N}\left[\frac{\sum_{j=1}^{n} f''_{ij} D_{hj}}{D}\right]_{O_k} = 0 \qquad (i \neq h;\, i,\ h = 1, 2, \ldots, n). \quad (40)$$

Taking into account (34) and (39), we can express (38) and (40) in the following compact form.

Let $f_h(x_1, x_2, \ldots, x_n; X_1, X_2, \ldots, X_n)$ $(h = 1, 2, \ldots, n)$ *be* n *integral rational functions with degrees respectively* a, b *in the two sets of* n *variables* x, X, *such that the* n *algebraic equations* (33) *are all of degree* $a + b$ *in the* x*'s. If these equations have* $(a+b)^n$ *simple proper solutions in common, given by*

$$x_1 = c_{1k},\ x_2 = c_{2k}, \ldots, x_n = c_{nk} \quad (k = 1, 2, \ldots, (a+b)^n),$$

then the following matrix equation holds

$$\sum_{k=1}^{(a+b)^n} \mathfrak{J}''_k\,(\mathfrak{J}'_k + \mathfrak{J}''_k)^{-1} = b\,(a+b)^{n-1}\,\mathfrak{J},$$

where $\mathfrak{J}$ *denotes the unit matrix of order* n, $\mathfrak{J}'$ *and* $\mathfrak{J}''$ *are the Jacobian matrices of the* f_h*'s with respect to the* x_i*'s and the* X_i*'s, and the index* k *signifies that the* x_i*'s,* X_i*'s must be replaced as follows*

$$x_1 = X_1 = c_{1k},\ x_2 = X_2 = c_{2k}, \ldots, x_n = X_n = c_{nk}.$$

Historical Notes and Bibliography

For a previous exposition of the various proofs of the theorem of Jacobi and of its generalization, given by the first theorem of § 41, p. 90, cf. Netto [81, §§ 461—465]. The present simple proof of this generalization in § 41 is taken from B. Segre [113], where is also to be found a further extension — also obtained here (p. 90) — relative to the case in which the r hypersurfaces of S_r have a V_d in common. The special case of the final result, when $r = 3$, $d = 1$, had already been obtained, by a different method and with some unnecessary restrictions, by End [44].

The formulae (3) and (4) of § 42, p. 91, are to be found in B. Segre [113], and equation (5) in Bompiani [24]. Historically, the first extensions of the theorem of Reiss to ordinary space are those of Bompiani [26] and, particularly, [27]. In the latter paper the final theorem of § 42, p. 93, already appears, but there it is limited to the investigation — by algebraic means — of the case in which second order elements with collinear centres lie in an ambient space of three dimensions.

The method used here — again that of B. SEGRE [113] — could yield further extensions, as e. g. suggested in the final paragraph of § 42, p. 93, and in the antepenultimate paragraph of § 47, p. 103, but such extensions still remain to be developed.

The results of §§ 43—49, pp. 93—107, appear here for the first time extended so far; however, some of them were already summarized (without proofs) in B. SEGRE [108]. We have here been dealing with a field in which there is still much to be done, as can be inferred from the preceding pages.

Part Seven

Veronese Varieties and Modules of Algebraic Forms

In the present Part we shall study some topologico-differential properties of varieties satisfying certain differentiability conditions. This work is a little separated from that exibited in the previous Parts, in view of the fresh methods which we employ. The local properties which we shall obtain can be used to establish a number of algebro-differential properties in the large. This fact we shall here demonstrate with particular reference to the VERONESE varieties, and to the question of representing an algebraic form as a linear combination of other forms.

§ 50. n-regular points of differentiable varieties. — Let P be a (simple) point of a d-dimensional differentiable variety W $(d \geqq 1)$, which has differential class C^n, where $n \geqq 2$. We shall be concerned with the *local topologico-differential properties* of W at such a point P; i. e., with those properties, possessed by sub-varieties of W passing through P, which are unaltered when the neighbourhood of P on W is subjected to a regular transformation which has differentiable class C^n. Consequently, if such a transformation transforms the neighbourhood of P into a neighbourhood of the homologue, O, of P on the transformed variety, V, then the above properties coincide with the analogous properties of V at O; in this case we shall therefore say that V constitutes a model of W.

We can always choose as a model for W (a region of) a projective space S_d. Such a choice naturally affords certain advantages, in so far as it simplifies our ideas. However, it will be convenient — for many questions — to take as our model some variety $V = V_d$ which is *n-regular* at O, i. e. such that its n-osculating space at O has the regular dimension, r, given by

$$r = \binom{n+d}{d} - 1 . \tag{1}$$

If we consider a variety V_d of differential class C^n immersed in a projective space S_N $(N \geqq r)$, with coordinates $x_0, x_1, \ldots, x_N$, then V_d can be represented parametrically in the form

$$x_i = x_i(u_1, u_2, \ldots, u_d) \qquad (i = 0, 1, \ldots, N),$$

where $x_i(u_1, u_2, \ldots, u_d)$ is a function of the parameters $u_1, u_2, \ldots, u_d$, and is differentiable n times with continuous derivatives with respect to them. It is not restrictive to suppose, as we shall, that the simple point O is the point given by the vanishing of all the parameters. It is evident that V_d is n-regular at its generic point if, and only if, the sets of partial derivatives $\frac{\partial^d x_i}{\partial u_{j_1} \partial u_{j_2} \cdots \partial u_{j_d}}$ $(i=0, 1, \ldots, N;\ s=0, 1, \ldots, n)$ are linearly independent, i. e. the x_i's satisfy no linear partial differential equations with order $\leqq n$.

Let us now fix such an n-regular model V_d, and let us choose on it a regular curvilinear *differential element*, E_h, with order h $(< n)$ and centre O. We can, with every integer k satisfying

$$0 \leqq h < k \leqq n,$$

associate an *algebraic variety*, $\{h, k\}$, of dimension $(k-h)d + h$, generated by the $\infty^{(k-h)(d-1)}$ spaces S_k which osculate at O the various regular curvilinear elements E_k lying on V_d and containing E_h. We note two particular cases of these varieties.

a) — If $h = 0$, E_h reduces to the point O, so that the variety $\{0, k\}$ is uniquely defined by the point O on V_d. The ambient of this variety is none other than the k-osculating space to V_d at O, i. e., the space joining the point O and its derivate points up to those of order k; in particular, $\{0, 1\}$ obviously concides with the tangent d-space to V_d at O.

b) — The variety $\{h, h+1\}$ reduces simply to a linear space S_{d+h}, and its generating spaces $S_k = S_{h+1}$ constitute an ∞^{d-1} star with centre S_h, the osculating space of E_h at O. If $h \geqq 1$, the above S_{d+h} contains as a hyperplane the S_{d+h-1} which joins S_h to the tangent space S_d (the 1-osculating space) of V_d at O, the intersection of S_h and S_d being simply the tangent line to E_h at O.

In general we can, moreover, prove that:

If the model V_d is changed arbitrarily, each variety $\{h, k\}$ — considered as being generated by $\infty^{(k-h)(d-1)}$ spaces S_k — is subjected only to a birational transformation. In particular in the case b), i. e. when $k = h + 1$, and if moreover $h > 1$, this transformation is a homography, operating on the ∞^{d-1} star to which $\{h, k\}$ is reduced; also, the homography transforms the hyperplane S_{d+h-1}, mentioned above, into the analogous hyperplane of the transformed star.

It is now evident that, given on V_d a certain number of elements with centre O, in order to obtain their topologico-differential invariants, it suffices to determine the invariants over certain birational transformations (or, in particular, over some homographies) of the varieties $\{h, k\}$ which are associated with those elements and their sub-elements E_h; it may, however, also be necessary to determine further geometric covariants of the given elements.

We obtain a very simple example by considering on V_d $d+1$ elements E_{h+1} containing the same E_h, where $h \geqq 1$. Recalling the preceding remarks concerning the case b), we see that, in present circumstances, it is permissible to represent the E_{h+1}'s as $d+1$ points of an affine S_{d-1}. Since such a set of points admits in general $d-1$ independent invariants, and no more, we deduce that:

On a differentiable V_d of class C^n, where $d \geqq 2$, $n \geqq h+1 \geqq 2$, a general set of $d+1$ regular curvilinear differential elements of order $h+1$, having in common an element of order h, admits $d-1$ independent topological invariants, and no more.

Before passing to a more detailed study of n-regular points of varieties, we remark that considerations of the above type can often be simplified by a suitable choice of the model V_d. For example, it is always possible to choose a VERONESE *variety* $V_d^{(m)}$ as such a model, i. e., the variety representing the linear system formed by all the algebraic forms of order m which belong to a space S_d. In § 56, p. 122, we shall see that, for $V_d^{(m)}$ to be n-regular at any point, it is necessary and sufficient that $m \geqq n$; thus we obtain the simplest model (with the least order) by taking $m = n$. We shall return to the consideration of such models in §§ 56—61, pp. 121—131.

§ 51. Some special properties of n-regular points of differentiable varieties. — We shall first prove the following theorem.

If a differentiable V_d is n-regular at one of its points, O $(d \geqq 1, n \geqq 1)$, and if p is any integer satisfying $2 \leqq p \leqq n+1$, then there exists a neighbourhood of O on V_d such that p arbitrary distinct points of that neighbourhood are always linearly independent. Further, if such a set of p points tend to O arbitrarily, then each S_{p-1}^0 of accumulation of the S_{p-1} joining the points lies entirely in the $(p-1)$-osculating space of V_d at the point O.

In the neighbourhood of O on V_d, we choose, in any manner, a system of admissible internal coordinates $(u_1, u_2, \ldots, u_d)$ which, without restriction, we may suppose to be all zero at O. If we define the integer r by means of (1), then, in virtue of the n-regularity of V_d at O, we see that the ambient S_ϱ of V_d has dimension $\varrho \geqq r$. Let us introduce in S_ϱ projective non-homogeneous coordinates $(x_1, x_2, \ldots, x_\varrho)$, with O as origin; we shall, when necessary for reasons of symmetry, also take $x_0 = 1$ as a $(\varrho+1)$-th homogeneous coordinate. The coordinates $x_1, x_2, \ldots, x_\varrho$ are then functions of $u_1, u_2, \ldots, u_d$ of class C^n on V_d, which all vanish at O; they can therefore — if we apply the TAYLOR expansion finishing at terms of degree n — be written, to within infinitesimals of degree $> n$ in the u_i's, as polynomials of degree $\leqq n$ in the u_i's without constant terms. In view of our hypotheses for V_d, there are precisely r of these polynomials which are linearly independent. Hence, by applying a suitable linear homogeneous substitution with constant coefficients on the coordinates, or rather by means of a suitable choice of coordinates, we can reduce the equations of V_d to the form

$$\begin{cases} x_i \overset{n}{=} \alpha_i & \text{for } i = 0, 1, \ldots, r, \\ x_i \overset{n}{=} 0 & \text{for } i = r+1, r+2, \ldots, \varrho, \end{cases} \tag{2}$$

where the sign $\overset{n}{=}$ signifies equality to within infinitesimals of order greater than n in the u_j's, and the α_i's are the $r+1$ monomials of degree not greater than n in the u_j's, i. e.

$$\alpha_i = u_1^{i_1} u_2^{i_2} \ldots u_d^{i_d} \qquad (0 \leq i_1 + i_2 + \cdots + i_d \leq n), \tag{3}$$

ordered — in a fixed manner — for non-decreasing values of the degree $i_1 + i_2 + \cdots + i_d$, so that, in particular, $\alpha_0 = 1$. The second set of equations (2) has, of course, to be omitted if $r = \varrho$.

Let $P_1, P_2, \ldots, P_p$ be p distinct points of V_d, which are sufficiently near to O. The points P_j (for $j = 1, 2, \ldots, p$) will be given e. g. by the following (sufficiently small) values of u:

$$u_1 = u_{1j}, \; u_2 = u_{2j}, \; \ldots, \; u_d = u_{dj}; \tag{4}$$

and we shall suppose that x_i assumes the corresponding value x_{ij}. In order to prove the first part of the theorem, it is necessary and sufficient to show that, however the points P_j are chosen in a suitable neighbourhood of O on V_d, *the matrix*

$$\|x_{ij}\|_{i=0,1,\ldots,\varrho;\, j=1,\ldots,p} \tag{5}$$

has rank p. Whence it is sufficient to prove that *p is the rank of the matrix*

$$\|x_{ij}\|_{i=0,\ldots,t;\, j=1,\ldots,p} \tag{6}$$

in that neighbourhood, where, for brevity, we have put

$$t = \binom{p+d-1}{d} - 1\,. \tag{7}$$

In order to obtain a contradiction, we suppose that, in fact, for every $\lambda > 0$ we can determine, in the neighbourhood of O with radius λ, p distinct points $P_j^{(\lambda)}(x_{0j}^{(\lambda)}, x_{1j}^{(\lambda)}, \ldots, x_{\varrho j}^{(\lambda)})$, such that

$$\det \|x_{ij}^{(\lambda)}\|_{i=i_1, i_2, \ldots, i_p;\, j=1,2,\ldots,p} = 0 \tag{8}$$

for each combination $(i_1, i_2, \ldots, i_p)$ of p integers from the set $0, 1, \ldots, t$.

If we define $\alpha_{ij}^{(\lambda)}$ in the manner suggested by (3), then we see from (2) that

$$\det \|x_{ij}^{(\lambda)}\|_{\substack{i=i_1, i_2, \ldots, i_p \\ j=1,2,\ldots,p}} = (1 + \varepsilon_{i_1, i_2, \ldots, i_p}^{(\lambda)}) \det \|\alpha_{ij}^{(\lambda)}\|_{\substack{i=i_1, i_2, \ldots, i_p \\ j=1,2,\ldots,p}},$$

where the ε's tend to zero with λ. Hence (8) implies that, for each combination $(i_1, i_2, \ldots, i_p)$,

$$\det \|\alpha_{ij}^{(\lambda)}\|_{i=i_1, i_2, \ldots, i_p; j=1,2,\ldots,p} = 0 \tag{9}$$

for an infinite number of possible values of λ. Thus, since the number of combinations is finite, there exist some values of λ for which (9) holds relative to every combination, i. e. for which the matrix

$$\|\alpha_{ij}^{(\lambda)}\|_{i=0,1,\ldots,t; j=1,2,\ldots,p}$$

has rank less than p.

Consequently, among the p equations

$$\sum_{i=0}^{t} c_i \alpha_{ij}^{(\lambda)} = 0 \qquad (j = 1, 2, \ldots, p) \tag{10}$$

in the $t+1$ unknowns $c_0, c_1, \ldots, c_t$ there is certainly one equation which is a consequence of the remaining $p-1$. But, from (3) and (7), we see that equations (10) express the conditions that the general polynomial of degree $p-1$ in the u's,

$$f(u) = \sum_{i_1+i_2+\cdots+i_d=0}^{p-1} c_i\, u_1^{i_1} u_2^{i_2} \ldots u_d^{i_d},$$

vanishes when we replace $u_1, u_2, \ldots, u_d$ by the corresponding internal coordinates (4) of the point $P_j^{(\lambda)}$ $(j = 1, 2, \ldots, p)$. Hence one of these conditions is a consequence of the remaining $p-1$. But this obviously does not occur for the polynomial

$$f(u) = \Phi_1(u)\, \Phi_2(u) \cdots \Phi_{p-1}(u),$$

where $\Phi_i(u)$ is a linear non-homogeneous polynomial in the u_i's which vanishes at $P_{j_i}^{(\lambda)}$ but not at $P_{j_1}^{(\lambda)}, \ldots, P_{j_{i-1}}^{(\lambda)}, P_{j_{i+1}}^{(\lambda)}, \ldots, P_{j_{p-1}}^{(\lambda)}$ $(i = 1, 2, \ldots, p-1)$, the integers $j_1, j_2, \ldots, j_{p-1}$ being all different.

Thus a contradiction is obtained and the first part of the theorem is proved.

Using this result, we see that p distinct points $P_1, P_2, \ldots, P_p$ lying on V_d, and sufficiently near to O, determine a S_{p-1}. This space has as its Grassmann coordinates the various minors, $p_{i_1, i_2, \ldots, i_p}$, of order p extracted from the matrix (5), where $(i_1, i_2, \ldots, i_p)$ is a combination of p integers of the set $0, 1, \ldots, \varrho$. We have seen that the minors, $q_{i_1, i_2, \ldots, i_p}$, of order p extracted from the matrix (6) are certainly not all zero; but these minors are merely the $p_{i_1, i_2, \ldots, i_p}$'s for which $i_1, i_2, \ldots, i_p$ have the values $0, 1, \ldots, t$.

We now suppose that the points $P_1, P_2, \ldots, P_p$ tend simultaneously to O, in an arbitrary manner, whilst they remain distinct. If S_{p-1}^0 is any element of accumulation of the S_{p-1} which joins the points, then we can always suitably restrict the manner in which the points tend to O, such that there exists the limiting position of S_{p-1}, given by S^0_{p-1}. Moreover, possibly with further restriction on the way in which the points tend to O, we can ensure that a suitable coordinate $q_{i_1, i_2, \ldots, i_p}$ is always non-zero so long as the points $P_1, P_2, \ldots, P_p$ are distinct (the limit of such a coordinate will of course be zero); whilst the ratios of the remaining coordinates to this one admit finite limits.

Using equations (2) and (3) we now see that, as $P_1, P_2, \ldots, P_p$ tend to O in the manner just specified, each $p_{i_1, i_2, \ldots, i_p}$, which is not also a q, tends to zero with an order of infinitesimal greater than unity relative to at least one q. Whence we can observe that

The Grassmann coordinates $p_{i_1, i_2, \ldots, i_p}$ of S^0_{p-1}, which are not coordinates q, are all zero.

The remaining coordinates of S^0_{p-1} are precisely unity and the finite limits of the ratios mentioned above.

This result is evidently equivalent to saying that S^0_{p-1} must lie in the space, S_t, given in S_ϱ by the equations

$$x_{t+1} = x_{t+2} = \cdots = x_\varrho = 0.$$

But, from (2), (3) and (7), we see that this space S_t is simply the $(p-1)$-osculating space to V_d at O. We have now proved the remainder of the theorem enunciated at the beginning of this section.

We note that there remains the question of deciding on what hypotheses, if any, the space S^0_{p-1} must necessarily belong to the variety $\{0, p-1\}$ relative to the point O of V_d.

We shall show, by example, that *the variety V_d no longer possesses the properties stated in the above theorem, when we remove the hypothesis of the n-regularity of V_d at O.* But first we prove a corollary of the above theorem.

Let V_d be cut by an S_k in a curve, or more generally, in an infinite set of points, $\mathfrak{J}$, for which O is a point of accumulation. In every neighbourhood of O on V_d, there is then a $(k+2)$-ple of linearly dependent distinct points, given, for example by $k+2$ distinct points of $\mathfrak{J}$ conveniently close to O. On the hypotheses of the theorem, $k+2 = p \leqq n+1$ cannot hold; hence $n \leqq k$. In other words, we have:

If V is n-regular at one of its points, O $(n \geqq 2)$, and if S_k is a space passing through O and having dimension $k < n$, then the intersection of V and S_k contains O as an isolated point.

Let $\mathfrak{C}$ be a differentiable curve of S_3, with differential class C^n, and let $\mathfrak{C}$ have an inflexion at the simple point O. There then exist in S_3 planes near to the tangent to $\mathfrak{C}$ at O, which contain three distinct points of $\mathfrak{C}$ near to O, but which are not proximate to the osculating plane of $\mathfrak{C}$ at O. This is not a contradiction of our theorem above, since $\mathfrak{C}$ is not 2-regular at O.

We now imagine the S_3 immersed in an S_ϱ, $\varrho > 3$, and we consider a differentiable V_d generated by ∞^1 S_{d-1}'s of S_ϱ which pass through the separate points of $\mathfrak{C}$ $(2 \leq d < \varrho)$. It is easily seen that the 2-osculating space to V_d at O is just the space which joins the three generators S_{d-1} passing through O or one of the two points succeeding O on $\mathfrak{C}$. This space then has dimension not exceeding $3d-2$. Thus V_d cannot possibly be 2-regular at O, and it is hardly surprising that the second part of the above theorem does not hold for $p = 3$, $n = 2$. Indeed, there manifestly exist planes of accumulation for planes joining three points of V_d near to O, which lie in S_3, without, however, lying in the 2-osculating space of V_d at O.

§ 52. On the freedom of hypersurfaces having assigned multiplicities at a set of points. — As was remarked in § 50, p. 110, the simplest models of differentiable varieties, for which the theorem of § 51, p. 110, holds, are the VERONESE varieties, $V_d^{(n)}$. In order to prepare the way for the study of these varieties, which appears in the final sections of this Part, we shall, in this and the following three sections, give some properties of algebraic forms and modules of algebraic forms. Many of these results are well known; but, at several points, we shall extend them so as to attain a certain completeness.

The totality of algebraic hypersurfaces of order n $(\geqq 1)$ in a given projective space, S_d $(d \geqq 1)$, constitutes an ∞^r linear system, where r is expressed by the relation (1). Thus the general form of degree n in $d+1$ variables is linearly (and homogeneously) dependent on

$$r + 1 = d_n$$

parameters, where, for brevity, we have put

$$d_n = \binom{d+n}{n} = \frac{(n+1)(n+2)\cdots(n+d)}{d!} \quad \text{if } n > 0 . \tag{11}$$

We shall extend, conventionally, the definition of d_n by assuming

$$d_0 = 1, \quad \text{and} \quad d_n = 0 \text{ if } n < 0: \tag{11'}$$

then the final expression for d_n in (11) holds (not only for positive values of n, but) also for

$$n = 0, -1, -2, \ldots, -d;$$

however, it holds for no other negative values of n.

For the hypersurfaces of order n in S_d, the passage through a given point with assigned multiplicity m, where $0 \leqq m \leqq n+1$, imposes d_{m-1} linearly independent conditions. The passing through $s \geqq 2$ distinct points $P_1, P_2, \ldots, P_s$ with multiplicities $m_1, m_2, \ldots, m_s$ respectively, thus imposes

$$d_{m_1-1} + d_{m_2-1} + \cdots + d_{m_s-1} \tag{12}$$

linear conditions; these may, however, be mutually dependent. But they are certainly independent, if

$$n \geqq m_1 + m_2 + \cdots + m_s - 1 \tag{13}$$

(cf. BERTINI [13], pp. 288—290).

With the hypothesis $s = 2$, the above conditions are, on the other hand, dependent when (13) does not hold; i. e., when

$$m_1 + m_2 \geqq n + 2. \tag{14}$$

More precisely, *in this case,* $d_{m_1+m_2-n-2}$ *of the linear conditions are a consequence of the remainder.* This is seen immediately when we assume the points P_1, P_2 to be $(1, 0, 0, \ldots, 0)$ and $(0, 1, 0, \ldots, 0)$ respectively, taking the homogeneous coordinates of a point of S_d to be $(x_1, x_2, \ldots, x_{d+1})$. Then P_1 and P_2 are points of multiplicity m_1 and m_2 for a hypersurface of order n if, and only if, in the relative equation, the monomial terms containing $x_1^{n-m_1+1}$ and $x_2^{n-m_2+1}$ as factors are missing. Thus the conditions imposed are distinct apart from those which concern the terms with $x_1^{n-m_1+1}\, x_2^{n-m_2+1}$ as a factor; and so the condition stating that such a term should be missing is counted precisely twice in the set of conditions. Removing this monomial factor from the single terms of the general form of degree n which contain it, we obtain the single terms of the general form of degree

$$n - (n - m_1 + 1) - (n - m_2 + 1) = m_1 + m_2 - n - 2\,;$$

and this number is non-negative, since we have imposed condition (14). Moreover, the number of coefficients, $d_{m_1+m_2-n-2}$, appearing in the form last obtained is equal to the number of repeated conditions. Hence the assertion is proved.

The conditions imposed on the multiple points P_1 and P_2 imply that x_1 and x_2 can appear with the sum of their degrees at most $2n - m_1 - m_2$, if (14) holds; thus *a hypersurface of order n, which passes through P_1 and P_2 with multiplicities m_1 and m_2, must then necessarily contain the line $P_1 P_2$ with multiplicity not less than $m_1 + m_2 - n$.* This observation clarifies, in a geometric manner, the occurrence of the preceding result.

§ 53. On the effective dimension of certain linear systems of hypersurfaces. — We shall denote by capital letters forms in the $d + 1$ variables x_i and by the corresponding small letters their respective degrees. Let

$$K_1, K_2, \ldots, K_h$$

be h forms (with degrees $k_1, k_2, \ldots, k_h$); they define a *module* $(K_1, K_2, \ldots, K_h)$ in the sense of BERTINI ([13], p. 307; this is a more restrictive meaning than that of KRONECKER and HILBERT). *Module* will be used in this restrictive sense only for the remainder of Part 7. The module is precisely the totality of forms expressible by an identity of the type

$$N = A_1K_1 + A_2K_2 + \cdots + A_hK_h, \tag{15}$$

where, if n is the degree of N, A_i is a form of degree

$$a_i = n - k_i \qquad (i = 1, 2, \ldots, h).$$

We wish to calculate the number, $d^{(n)}_{k_1, k_2, \ldots, k_h}$, of essential parameters, on which N depends for the various A_i, on the hypotheses that

$$1 \leqq h \leqq d + 1$$

and that the intersection of the h hypersurfaces $K_1 = 0, K_2 = 0, \ldots, K_h = 0$ of S_d has regular dimension $d-h$. This means that the hypersurfaces have no point in common if $h = d + 1$, whilst for $h \leqq d$ the intersection variety is pure and of dimension $d - h$ (possibly having multiple components). Clearly, the number $d^{(n)}_{k_1, k_2, \ldots, k_h}$ diminished by unity is equal to the *dimension of the linear system*, $(K_1, K_2, \ldots, K_h)_n$, consisting of the forms of order n in the module $(K_1, K_2, \ldots, K_h)$. We shall prove the following theorem, which is already known for the case in which the above-mentioned intersection is without multiple components (cf. SEVERI [131], and BERTINI [13] pp. 311—315).

With the above hypotheses, and in the notation of § 52, *p.* 114,

$$d^{(n)}_{k_1, k_2, \ldots, k_h} = \Sigma\, d_{n-k_i} - \Sigma\, d_{n-k_i-k_j} + \cdots + \\ + (-1)^{h-1} d_{n-k_1-k_2-\cdots-k_h}, \tag{16}$$

where the summations on the right-hand side extend over the simple combinations of 1, 2, . . . *integers from* 1, 2, . . ., h.

If $h = 1$, then a form N given by (15) corresponds to precisely one A_1; and so the forms N depend on $d_{a_1} = d_{n-k_1}$ parameters, all of which are essential. This proves (16) for $h = 1$. We can therefore suppose that $h > 1$ and, in order to proceed by induction, that the theorem is proved for modules determined by less than h forms. We note that the h hypersurfaces have a regular intersection, only if the same happens for a subset of $1, 2, \ldots, h - 1$ hypersurfaces chosen arbitrarily from them.

On the right-hand side of (15), $\Sigma\, d_{a_i}$ $(= \Sigma\, d_{n-k_i})$ coefficients appear linearly and homogeneously, these being just the coefficients which appear in the A_i's. However, since $h > 1$, each N is derived from several choices of the A_i's; all these choices can be obtained from one of them in an obvious manner, when we have solved the diophantine equation

$$A_1K_1 + A_2K_2 + \cdots + A_hK_h = 0. \tag{17}$$

Hence, if c is the number of parameters on which the A_i's depend when they satisfy (17), we see that

$$d^{(n)}_{k_1, k_2, \ldots, k_d} = \Sigma\, d_{n-k_i} - c. \tag{18}$$

We recall (see BERTINI [13], pp. 329—330) that the most general solution of (17) is presently obtained by putting

$$A_i = \sum_{j=1}^{h} A_{ij} K_j \qquad (i = 1, 2, \ldots, h), \ (19)$$

where the A_{ij}'s are forms of degree

$$a_{ij} = n - k_i - k_j$$

in the x's, and where

$$A_{ij} = -A_{ji} \qquad (i, j = 1, 2, \ldots, h) \ (20)$$

(so that the A_{ii}'s are identically zero). The final equation of (19) simply states that A_h belongs to the module $(K_1, K_2, \ldots, K_{h-1})$. Therefore, from the inductive hypothesis, we deduce that A_h depends on $d^{(a_h)}_{k_1, k_2, \ldots, k_{h-1}}$ essential constants, where $d^{(a_h)}_{k_1, k_2, \ldots, k_{h-1}} = d^{(n-k_h)}_{k_1, k_2, \ldots, k_{h-1}}$ is given by a relation analogous to (16). For a given A_h, we can now suppose that we have fixed the A_{hj}'s; whence, by (20), we obtain the A_{jh}'s. Then, the penultimate relation of (19) expresses the fact that $A_{h-1} - A_{h-1,h} K_h$ belongs to the module $(K_1, K_2, \ldots, K_{h-2})$. Therefore, again from the inductive hypothesis, we deduce that A_{h-1} depends on $d^{(n-k_{h-1})}_{k_1, k_2, \ldots, k_{h-2}}$ essential constants.

Thus we can proceed, step by step, using equation (20) and considering the relations (19) in inverse order. We finally deduce, from the second relation of (19), that

$$A_2 - \sum_{j=3}^{h} A_{2j} K_j = A_2 + \sum_{j=3}^{h} A_{j2} K_j$$

belongs to the module (K_1). Whence A_2 depends on $d_{n-k_1-k_2} = d^{(n-k_2)}_{k_1}$ essential constants, which are simply the coefficients of A_{21}, to which we can give arbitrary values. Thus, from (20), A_{12} is now fixed; so that A_1 is uniquely determined by the first relation of (19). Consequently we obtain

$$c = d^{(n-k_h)}_{k_1, k_2, \ldots, k_{h-1}} + d^{(n-k_{h-1})}_{k_1, k_2, \ldots, k_{h-2}} + \cdots + d^{(n-k_2)}_{k_1}.$$

It now suffices to substitute this expression for c in (18), and to replace the symbols on the right-hand side by their values in the form (16), in order to deduce relation (16) itself.

We note that, in view of the preceding analysis, if the coefficients of the forms $K_1, K_2, \ldots, K_h$ are *rational functions of one or more parameters*, and if, for every choice of these parameters, N belongs to the module $(K_1, K_2, \ldots, K_h)$, then a relation of the type (15) holds, for which the coefficients of the forms $A_1, A_2, \ldots, A_h$ can be taken as *rational functions of the same parameters*.

§ 54. Two relations of Lasker concerning modules of hypersurfaces. — We shall now establish two theorems of LASKER [61], as simple consequences of the results of the preceding section (cf. also SEVERI [135], and BERTINI [13] pp. 324—328).

Firstly, we shall suppose that

$$n-(k_1+k_2+\cdots+k_h)\geqq -d; \tag{21}$$

then, from the second paragraph of § 52, we see that, in order to calculate each of the d_i's which appear on the right-hand side of (16), we can use the formula (11). Now, putting successively $h=d+1$, $d, d-1, \ldots$, we obtain, from (16),

$$d^{(n)}_{k_1,k_2,\ldots,k_{d+1}}=d_n, \tag{22}$$

$$d^{(n)}_{k_1,k_2,\ldots,k_d}=d_n-k_1k_2\ldots k_d, \tag{23}$$

$$d^{(n)}_{k_1,k_2,\ldots,k_{d-1}}=d_n-\tfrac{1}{2}k_1k_2\ldots k_{d-1}(2n-\Sigma k_i+d+1),$$

$$d^{(n)}_{k_1,k_2,\ldots,k_{d-2}}=d_n-\tfrac{1}{2}k_1k_2\ldots k_{d-2}\{\tfrac{1}{3}\Sigma k_i^2+\tfrac{1}{2}\Sigma k_ik_j-$$
$$-\tfrac{1}{2}(2n+d+1)\Sigma k_i+\tfrac{1}{12}(3d+2)(d+1)+n(n+d+1)\}$$

(cf. SEVERI [133]). Equation (22), which is valid for n satisfying (21), gives at once the first theorem of LASKER:

If $d+1$ hypersurfaces of S_d with orders $k_1, k_2, \ldots, k_{d+1}$ have no common points, then every hypersurface with order $n\geqq\Sigma k_i-d$ belongs to their module.

An immediate consequence of (16) is

$$d^{(n-k_1)}_{k_2,k_3,\ldots,k_h}=d_{n-k_1}+d^{(n)}_{k_2,k_3,\ldots,k_h}-d^{(n)}_{k_1,k_2,\ldots,k_h}. \tag{24}$$

The various terms on the right-hand side of (24) are the dimensions of the sub-modules consisting of the forms of order n in the modules (K_1), $(K_2, K_3, \ldots, K_h)$, $(K_1, K_2, \ldots, K_h)$. Since the last of those sub-modules is the module which joins the other two, the dimension of the intersection of the two first sub-modules must be given by the right-hand side (and thus also by the left-hand side) of (24). Whence we obtain immediately the second theorem of LASKER:

Let $K_1, K_2, \ldots, K_h$, A be $h+1$ forms in $d+1$ variables, where $2\leqq h\leqq d+1$, and let the variety given by the vanishing of the first h of these forms have regular dimension. If the product AK_1 belongs to the module defined by $K_2, K_3, \ldots, K_h$, then A also belongs to the module.

This theorem can also be obtained as a particular case (in which $l=1$) of the following proposition, proved independently below.

If $1\leqq l<h\leqq d+1$, and if $K_1, K_2, \ldots, K_h$ denote h forms in $d+1$ variables such that the variety defined by the vanishing of them has regular dimension, then a necessary and sufficient condition that a form N in the same variables should belong simultaneously to the modules $(K_1, K_2, \ldots, K_l)$ and $(K_{l+1}, K_{l+2}, \ldots, K_h)$ is that it should belong to the module defined by the $l(h-l)$ products K_iK_j $(i=1,2,\ldots,l;\ j=l+1,l+2,\ldots,h)$.

First of all, we remark that the sufficiency of the condition enunciated is evident; thus we need only to establish its necessity.

We suppose, with the obvious meaning for the symbols, that N can be written in the following two ways

$$N = \sum_{i=1}^{l} A_i K_i, \qquad N = \sum_{j=l+1}^{h} (-A_j) K_j.$$

Whence we obtain equation (17); and therefore also (19) and (20). Consequently, we now have:

$$N = \sum_{i=1}^{l} A_i K_i = \sum_{i=1}^{l} \sum_{j=1}^{h} A_{ij} K_i K_j = \sum_{i=1}^{l} \sum_{j=1}^{l} A_{ij} K_i K_j + \sum_{i=1}^{l} \sum_{j=l+1}^{h} A_{ij} K_i K_j;$$

but in the final expression the first summation is identically zero, in view of (20). Hence the assertion is proved.

§ 55. Some important criteria for a hypersurface to belong to a given module. — We first recall a theorem of SEVERI [131] (cf. also BERTINI [13], pp. 311—318).

If h hypersurfaces $K_1, K_2, \ldots, K_h$ of S_d $(1 \leqq h \leqq d)$ meet in a variety, M, which is simple (i. e. without multiple components) and of regular dimension $d-h$, then a necessary and sufficient condition for a hypersurface N of S_d to belong to the module $(K_1, K_2, \ldots, K_h)$ is that N should pass through M.

With the hypotheses of this theorem, we see, from § 53, p. 116, that the effective dimension of the linear system $|N|$ of hypersurfaces, N, which contain M, is precisely equal to $d^{(n)}_{k_1, k_2, \ldots, k_h} - 1$, where $d^{(n)}_{k_1, k_2, \ldots, k_h}$ is given by (16) taken together with the definition (11) and (11′) of d_n. In order to obtain the virtual dimension of the above linear system $|N|$, it is sufficient to modify the expression obtained in this manner, so that each one of the integers d_i's (for every integer i) — which appears in the expression — is calculated according to the formula (11), n being replaced here by i; this means that the dimension should be calculated without taking into account (11′). It is now therefore possible to calculate the number σ which is the difference between those two dimensions, this number being the so-called superabundance of the system $|N|$. For example, if we assume that n has the greatest value which does not satisfy (21), i. e. if

$$n = k_1 + k_2 + \cdots + k_h - d - 1,$$

then applying (11) or (11′) we get $d_{-d-1} = (-1)^d$ or $d_{-d-1} = 0$ respectively; whence at present we infer

$$\sigma = (-1)^{h+d}.$$

In particular, for $h = d$, we obtain the following well-known result (cf., for example, B. SEGRE [113], nn. 1, 14).

d general hypersurfaces of S_d, with orders $k_1, k_2, \ldots, k_d$, meet in $k_1 k_2 \ldots k_d$ points which impose only $k_1 k_2 \ldots k_d - 1$ distinct conditions on the hypersurfaces of order $k_1 + k_2 + \cdots + k_d - d - 1$ containing them.

We now recall that h hypersurfaces of S_d $(2 \leqq h \leqq d)$ are said to give rise to the simple case at a common point P, if the intersection of the corresponding tangent cones, to the hypersurfaces at P, has the regular dimension. In the case of $h = d$, this is equivalent to saying that the hypersurfaces have no common tangent at P. Making use of this concept, we can formulate as below the following theorem of TORELLI ([160], n. 1; see also BERTINI [13], pp. 337—339). In the special case of $d = 2$, this result reduces to the classical «theorem $Af + B\varphi$» of NOETHER (see, for example, SEVERI [138], pp. 335—340) and, in the rather more general case of $h = d$, for any integer d, it becomes the extension of this classical theorem due to KÖNIG ([60], p. 389).

Let $K_1, K_2, \ldots, K_h$ be h hypersurfaces of S_d $(2 \leqq h \leqq d)$, which intersect in a variety, M, with regular dimension $d - h$, such that at the generic point of each component of M the simple case arises. Then, if $t_1, t_2, \ldots, t_h$ are the multiplicities of $K_1, K_2, \ldots, K_h$ at such a generic point, every hypersurface of S_d which contains the separate components of M with multiplicity greater than the corresponding $t_1 + t_2 + \cdots + t_h - h$ belongs to the module $(K_1, K_2, \ldots, K_h)$.

We shall establish this property in § 61, p. 130, by a new procedure of synthetic character. In that section we shall also prove other results, relative to the more general situation in which the hypersurfaces $K_1, K_2, \ldots, K_h$ intersect regularly, but do not necessary give rise to the simple case.

We shall now conclude this section by proving the following proposition, which is to be found in TORELLI [160] n. 3, and BERTINI [13] pp. 337—338. However, the proof in these two works is rather different from that given here.

Let $K_1, K_2, \ldots, K_h$ be h hypersurfaces of S_d $(2 \leqq h \leqq d-1)$ which intersect in a variety, M, with regular dimension $d - h$; and let N be a hypersurface such that its section by a generic S_l of S_d, for a fixed integer l satisfying $h \leqq l \leqq d - 1$, belongs to the module determined by the intersections of $K_1, K_2, \ldots, K_h$ with S_l. Then N belongs necessarily to the module $(K_1, K_2, \ldots, K_h)$.

Let us introduce in S_d homogeneous coordinates $(x_1, x_2, \ldots, x_{d+1})$, such that no component of M lies in a $(d-1)$-dimensional cone with the space S_{l-1}:

$$x_{l+1} = x_{l+2} = \cdots = x_{d+1} = 0$$

as vertex. This is certainly possible, since $l \geqq h$. We use the hypothesis of the theorem, with regard to the generic S_l of S_d which contains this S_{l-1}, taking as equations of such a space:

$$x_{l+2} = c_{l+2} x_{l+1}, \; x_{l+3} = c_{l+3} x_{l+1}, \; \ldots, \; x_{d+1} = c_{d+1} x_{l+1}, \tag{25}$$

where the ratios $1 : c_{l+2} : c_{l+3} : \cdots : c_{d+1}$ are fixed generically. If we replace $x_{l+2}, x_{l+3}, \ldots, x_{d+1}$ in each of $K_1, K_2, \ldots, K_h, N$ by means of (25), obtaining forms $K_1', K_2', \ldots, K_h', N'$ in $x_1, x_2, \ldots, x_{l+1}$, then there results an equality of the type

$$N' = A_1' K_1' + A_2' K_2' + \cdots + A_h' K_h', \tag{26}$$

where $A_1', A_2', \ldots, A_h'$ are forms in $x_1, x_2, \ldots, x_{l+1}$. The coefficients of $K_1', K_2', \ldots, K_h', N'$ are rational functions in the c_j's; thus it is not restrictive to suppose that the same property holds for the coefficients of the A_i's, in view of the final observation made in § 53, p. 117. We can now eliminate the c_j's from (26), by means of (25); and, multiplying each side of the relation so obtained by a suitable form, P, in $x_{l+1}, x_{l+2}, \ldots, x_h$, we see that the product PN belongs to the module $(K_1, K_2, \ldots, K_h)$. We notice here that the hypersurface $P = 0$ is a cone with vertex S_{l-1}, and therefore, from the definition of S_{l-1}, it cuts M in a variety with regular dimension $d - h - 1$. As a consequence of the second theorem of Lasker (§ 54, p. 118), it now follows that N must in fact belong to the module $(K_1, K_2, \ldots, K_h)$.

We remark that we have, in the course of the above proof, established a rather stronger result than that enunciated. We have, in fact, proved that:

The above proposition is valid if the properties enunciated, relative to the generic S_l, are possessed simply by the ∞^{d-l} spaces S_l of S_h which pass through a general S_{l-1} of S_d.

§ 56. Some properties of the osculating spaces at the points of a Veronese variety $V_d^{(n)}$. — As we have said towards the end of § 50, p. 110, we denote by $V_d^{(n)}$ the Veronese variety which represents the linear system of all hypersurfaces of order n lying in S_d. Such a variety is b i r a t i o n a l, having invariantive order equal to unity, with dimension d and order n^d. It lies in a space S_r of dimension $r = d_n - 1$, where r and d_n are given by (1) and (11) respectively. If we consider $u_0, u_1, \ldots, u_d$ as homogeneous coordinates in S_d, then the variety $V_d^{(n)}$ can be represented parametrically by

$$X_{i_0 i_1 \cdots i_d} = u_0^{i_0} u_1^{i_1} \ldots u_d^{i_d} \qquad (i_0 + i_1 + \cdots + i_d = n),$$

where the $X_{i_0 i_1 \cdots i_d}$'s are homogeneous coordinates in the S_r in which $V_d^{(n)}$ lies. From this representation, we see that the order of $V_d^{(n)}$ is equal in fact to the number, n^d, of intersections of d general hypersurfaces of order n in S_d. The homographies of S_d evidently induce on S_r a group of homographies which leaves $V_d^{(n)}$ invariant, and which acts transitively on $V_d^{(n)}$, i. e., there is always at least one such homography of S_r which transforms any point of $V_d^{(n)}$ into any other point of the variety. Thus we see that $V_d^{(n)}$ is free from singular points; and indeed, by putting $u_0 = 1$, we deduce, as in § 51, p. 111, that $V_p^{(n)}$ is n-regular at the point with coordinates $X_{n0\cdots 0} = 1$, $X_{i_0 i_1 \cdots i_d} = 0$ for $i_0 \neq n$; whence it follows that $V_d^{(n)}$ is n-regular at each of its points. The n-osculating space, at any point of the variety, thus coincides with the ambient space S_r.

In virtue of the definition of $V_d^{(n)}$, the different sub-spaces of S_r, considered as the centres of stars of hyperplanes, are in one-one correspondence with the various linear systems of hypersurfaces of order n lying in S_d; this correspondence is in fact homographic. Thus a sub-space S_k of S_r is associated with precisely one ∞^{r-k-1} linear system. In particular, if q denotes any integer satisfying $0 \leqq q < n$, we deduce that:

The q-osculating space, $S_{d_q - 1}$, to $V_d^{(n)}$ at any of its points, O, is the model of the $(r - d_q)$-dimensional linear system formed by the hypersurfaces of order n in S_d which contain the image point P of O with multiplicity not less than $q+1$.

Indeed, the hyperplanes of S_r which contain the above $S_{d_q - 1}$ have the characteristic property of cutting $V_d^{(n)}$ in a variety having a point of multiplicity not less than $q+1$ at O. Hence they correspond to those hypersurfaces of order n in S_d which pass through P with multiplicity not less than $q+1$.

We shall now interpret the results appearing towards the end of § 52, as geometric properties of the osculating varieties at points of $V_d^{(n)}$. In fact, if we put $m_i = q_i + 1$, we deduce immediately that:

If O_i $(i = 1, 2, \ldots, s)$ are s $(\geqq 2)$ distinct points of $V_d^{(n)}$, then the corresponding q_i-osculating space to $V_d^{(n)}$ at these points are linearly independent, when, besides the relation $0 \leqq q_i < n$, the following inequality holds:

$$n \geqq q_1 + q_2 + \cdots + q_s + s - 1 . \tag{27}$$

In particular, when $s = n+1$ and $q_1 = q_2 = \cdots = q_s = 0$, we obtain, in conformity with the theorem of § 51, p. 110, that *any $n+1$ distinct points of $V_d^{(n)}$ are always linearly independent.* And further:

If $n/2 \leqq q < n$, then any two distinct q-osculating spaces to $V_d^{(n)}$ are incident in a space of (non negative) dimension $d_{2q-n} - 1$. More generally, if q_1, q_2 $(< n)$ are two integers satisfying

$$q_1 + q_2 \geqq n,$$

then the spaces which are q_1-osculating and q_2-osculating to $V_d^{(n)}$, at two distinct points O_1 and O_2 respectively, intersect in a space of dimension $d_{q_1+q_2-n} - 1$; they are thus joined by a space of dimension $d_{q_1} + d_{q_2} - d_{q_1+q_2-n} - 1$. The latter space contains the different spaces which are $(q_1 + q_2 - n + 2)$-osculating to $V_d^{(n)}$ at the various points of the rational normal C^n lying on $V_d^{(n)}$ and passing through O_1 and O_2.

The C^n, which appears in the above result, is simply the image on $V_d^{(n)}$ of the line which joins the images of O_1 and O_2 in S_d.

§ 57. The ambients of certain subvarieties of $\mathbf{V}_d^{(n)}$. — The study of the algebraic subvarieties of $V_d^{(n)}$ is equivalent to the study of the algebraic varieties of an S_d: this is immediately derived from the way in which $V_d^{(n)}$ can be represented in S_d (§ 56, p. 121). However, as we shall see, many properties of such varieties are more easily enunciated and proved in the ambient, S_r, of $V_d^{(n)}$ than in S_d.

It is evident that an algebraic variety, K, of S_d is reducible or irreducible, pure or impure if, and only if, the same property is possessed by its image, $\varkappa$, in $V_d^{(n)}$. In the case in which K and $\varkappa$ are both pure, they have the same dimension, which we shall denote by h.

If the order of K is k, we see that the order of $\varkappa$ is kn^h, since this is the number of points which K has in common with h general hypersurfaces of order n in S_d. Further, if a point P of S_d is simple or multiple on K, then its image is respectively simple or multiple on $\varkappa$; in the latter case, the «composition» of the singularity of K at P is the same as that of $\varkappa$ at O. Nevertheless, the substitution of K by $\varkappa$ can be used in the study of singularities (cf. B. SEGRE [114]). Also, from such a substitution, certain advantages derive with regard to questions of intersection and postulation. For example, it is clear that:

The postulation of K, for the hypersurfaces of order n in S_d, exceeds by precisely unity the dimension of the ambient of $\varkappa$.

More generally, in order to obtain the postulation for the hypersurfaces of order n in S_d of an algebraic variety, which consists of several components (possibly with different dimensions) counted with any multiplicities m, it is sufficient to consider the image of these components on $V_d^{(n)}$, and then the varieties generated by the $(m-1)$-osculating spaces to $V_d^{(n)}$ at the points of the images. Recalling the results of the preceding section, we then have that:

The required postulation exceeds by unity the dimension of the subspace of S_r which is the join of the ambients of the various varieties generated as above.

We could further extend these researches, by giving a concrete image of the "*comportamento*", relative to the base varieties of the linear systems (of hypersurfaces of order n in S_d), which was recently defined by SEVERI [145, 146]. For this, it would suffice to introduce considerations of suitable differential elements of $V_d^{(n)}$ and their osculating spaces, and to investigate the way in which the birational correspondences operate between the various varieties $V_d^{(n)}$ for different values of the integer n. However, for reasons of space, we shall not dwell on such extensions at present; in the sequel, we shall limit ourselves to the study of simple and multiple primals on a $V_d^{(n)}$, and to certain questions concerning their intersections in the regular case.

If K is an algebraic hypersurface of order k in S_d, then its image $\varkappa$ on $V_d^{(n)}$ is a pure variety, of dimension $d-1$ and order $k\,n^{d-1}$; and, if K is simple, $\varkappa$ *belongs to a space of dimension* $d_n - d_{n-k} - 1$, where the symbols are given by equations (11) and (11′) of § 52, p. 114. This follows immediately from the first proposition of this section.

When K is reducible, with components $K_1, K_2, \ldots, K_h$ counted with multiplicities $m_1, m_2, \ldots, m_h$, so that

$$k = m_1 k_1 + m_2 k_2 + \cdots + m_h k_h,$$

then the variety $\varkappa$ representing K on $V_d^{(n)}$ is also reducible, with components $\varkappa_1, \varkappa_2, \ldots, \varkappa_h$, say, which are again counted with multiplicities $m_1, m_2, \ldots, m_h$. We see, from the second proposition of this section, that in order to represent K in S_r it is convenient to replace the variety $\varkappa$ by the one, $\varkappa^*$, which has as (simple) components those generated by the $(m_i - 1)$-osculating spaces to $V_d^{(n)}$ on the points of $\varkappa_i$ $(i = 1, 2, \ldots, h)$. Indeed, since the postulation of K manifestly depends only on the order k of K, we have that:

The ambient space of $\varkappa^*$ *still has dimension* $d_n - d_{n-k} - 1$; *it reduces to the ambient of* $\varkappa$ *if* K *is simple, and is a single-valued everywhere continuous function of* $\varkappa$.

We remark that, when $\varkappa$ is reducible in the manner indicated and is endowed with at least one multiple component, then its ambient space has dimension less than $d_n - d_{n-k} - 1$, the dimension being in fact $d_n - d_{n-(k_1+k_2+\cdots+k_h)} - 1$.

We now apply the theorem of SEVERI (§ 55, p. 119), using the theorem of § 53, p. 116, and also the second result of the present section. We thus obtain the following extension of the above result concerning the dimension of the ambient of $\varkappa$ when this variety is simple.

Let $\varkappa_1, \varkappa_2, \ldots, \varkappa_h$ *be* h *hypersurfaces of* $V_d^{(n)}$ $(2 \leq h \leq d)$, *which intersect in a variety,* μ, *free from multiple components and having regular dimension* $d-h$. *Then the ambient space of* μ *is precisely the intersection of the ambients of* $\varkappa_1, \varkappa_2, \ldots, \varkappa_h$; *the dimension of the former is* $d_n - d^{(n)}_{k_1, k_2, \ldots, k_h} - 1$, *where the symbols are given by* (16), (11) *and* (11′), *and* k_i *is an integer such that* $k_i n^{d-1}$ *is the order of* $\varkappa_i$ $(i = 1, 2, \ldots, h)$. *Further, the order of* μ *is* $m = k_1 k_2 \ldots k_h n^{d-h}$.

The case in which the above variety μ has multiple components necessitates rather deeper considerations, but it can be studied as a limit of the simple case. This is precisely what we now propose to do. We commence with the simplest possibility in which $h = d$, and so μ becomes a set of points.

§ 58. The isolated multiple intersections of d primals on $V_d^{(n)}$. — Let $\varkappa_1, \varkappa_2, \ldots, \varkappa_d$ be d (≥ 1) hypersurfaces of $V_d^{(n)}$ (with orders respectively $k_1 n^{d-1}, k_2 n^{d-1}, \ldots, k_d n^{d-1}$) which meet in a set of points, μ; μ then consists of precisely $m = k_1 k_2 \ldots k_d$ points, each point being counted with the appropriate multiplicity. If we denote by $O_1, O_2, \ldots, O_s$ the distinct points in μ, and by $p_1, p_2, \ldots, p_s$ their respective (positive) multiplicities, then

$$p_1 + p_2 + \cdots + p_s = m = k_1 k_2 \ldots k_d. \tag{28}$$

It is clear that the m points of μ are distinct (i. e., $s = m$ and $p_1 = p_2 = \cdots = p_s = 1$) if, and only if, $\varkappa_1, \varkappa_2, \ldots, \varkappa_d$ are non-singular at each of their common points, and give rise to the simple case at each such point (§ 55, p. 120).

The following signification (dynamic, according to SEVERI) of the multiplicity p_i is well-known. If the varieties $\varkappa_1, \varkappa_2, \ldots, \varkappa_d$ are displaced a little in a general manner on $V_d^{(n)}$, then the new varieties $\bar{\varkappa}_1, \bar{\varkappa}_2, \ldots, \bar{\varkappa}_d$ intersect in a set $\bar{\mu}$ of m distinct points, of which precisely p_i lie in the neighbourhood of O_i (or rather tend to O_i as the $\bar{\varkappa}_j$'s tend to the $\varkappa_j$'s). Evidently, if $\varkappa_1, \varkappa_2, \ldots, \varkappa_d$ arise from the forms $K_1, K_2, \ldots, K_d$ of S_d, then $\bar{\varkappa}_1, \bar{\varkappa}_2, \ldots, \bar{\varkappa}_d$ similary arise from forms $\bar{K}_1, \bar{K}_2, \ldots, \bar{K}_d$, where $\bar{K}_j$ is proximate to K_j; and these two varieties have the same order k_j ($j = 1, 2, \ldots, d$). Since, from § 53, p. 116, the linear system $(\bar{K}_1, \bar{K}_2, \ldots, \bar{K}_d)_n$ and $(K_1, K_2, \ldots, K_d)_n$ have the same dimension, the first of the systems tends to the second as the $\bar{K}_j$'s tend to the K_j's. In view of (1), (21), (23) and (28), on the hypothesis that

$$n \geq k_1 + k_2 + \cdots + k_d - d, \tag{29}$$

the dimension of these linear systems is $r - m$. This then certainly occurs when, as we shall suppose,

$$n \geq k_1 k_2 \ldots k_d - 1, \tag{30}$$

because (29) is a consequence of (30).

The two linear systems just considered are represented in S_r by two subspaces; and the subspace corresponding to the first of these, is simply, in virtue of the theorem of SEVERI stated in § 55, p. 119, the space joining the points of $\bar{\mu}$. Suppose that (30) holds, and therefore *a fortiori* so does (29): we deduce (from § 56, p. 122) that the two subspaces have dimension $m - 1$, and we denote them by $\bar{S}_{m-1}$ and S_{m-1} respectively. We remark that S_{m-1} is none other than the intersection of the ambients of $\varkappa_1, \varkappa_2, \ldots, \varkappa_d$, on the hypothesis that these varieties are free from multiple components. Moreover, S_{m-1} contains the s points $O_1, O_2, \ldots, O_s$ of μ; but, when $m > s$ — thus only excluding the case in which $p_1 = p_2 = \cdots = p_s = 1$ — S_{m-1} is not the space joining these points. We shall now prove the following theorem.

With the hypothesis (30), however the varieties $\bar{\varkappa}$ are chosen tending to the varieties $\varkappa$, the space $\bar{S}_{p_i-1}$ which joins the p_i points of $\bar{\mu}$ in the neighbourhood of O_i ($i = 1, 2, \ldots, s$) admits a uniquely defined limiting position, S_{p_i-1}. The space S_{p_i-1} contains O_i and lies in the ($p_i - 1$)-osculating space to $V_d^{(n)}$ at O_i. Moreover, the above space S_{m-1}, i. e. the limiting position of $\bar{S}_{m-1}$, the join of the points of $\bar{\mu}$, is precisely the space which joins the s spaces S_{p_i-1}: thus the latter are always mutually linearly independent.

We first observe that, in virtue of (28) and (30), the relation we deduce from (27) by putting $q_i = p_i - 1$ ($i = 1, 2, \ldots, s$) is satisfied. Hence, in view of the second proposition of § 56, p. 122, the s spaces which are ($p_i - 1$)-osculating to $V_d^{(n)}$ at O_i, for $i = 1, 2, \ldots, s$, are linearly independent; the same is therefore true, a fortiori, for any set of s spaces chosen one in each of these osculating spaces. Further, from (28) and (30), for each $i = 1, 2, \ldots, s$ we have $p_i \leq n + 1$.

Now let the $\bar{\varkappa}$'s tend to the $\varkappa$'s in any manner. We recall that, for any such variation, the space $\bar{S}_{m-1}$, joining the m points of $\bar{\mu}$, admits S_{m-1} as its definite limit. The space $\bar{S}_{m-1}$ clearly contains each of the spaces $\bar{S}_{p_1-1}, \bar{S}_{p_2-1}, \ldots, \bar{S}_{p_s-1}$ defined in the enunciation of the theorem. Let $S_{p_1-1}, S_{p_2-1}, \ldots, S_{p_s-1}$ be any set of spaces of accumulation for the $\bar{S}_{p_i-1}$'s. Then each S_{p_i-1} must be in S_{m-1}; moreover, from § 51, p. 110, the space S_{p_i-1}, as it is a limiting position of $\bar{S}_{p_i-1}$, must pass through O_i and lie in the ($p_i - 1$)-osculating space to $V_d^{(n)}$ at O_i.

Whence the spaces $S_{p_1-1}, S_{p_2-1}, \ldots, S_{p_s-1}$ are linearly independent, that is, they are joined by a space of dimension $p_1 + p_2 + \cdots + p_s - 1 = m - 1$. This join lies in the space S_{m-1}, since each of the S_{p_i-1}'s lies in this space, and so must coincide with S_{m-1}. Further, it now emerges that S_{p_i-1} must be precisely the intersection of S_{m-1} and the ($p_i - 1$)-osculating space to $V_d^{(n)}$ at O_i; consequently, S_{p_i-1} is uniquely defined, i. e., S_{p_i-1} ($i = 1, 2, \ldots, s$) is independent of the way in which the $\bar{\varkappa}$'s tend to the $\varkappa$'s. The proof is now complete.

§ 59. The regular multiple intersections on $V_d^{(n)}$. — We shall now extend the considerations of the preceding section to the case of the regular intersections of h ($< d$) primals on $V_d^{(n)}$.

Let $\varkappa_1, \varkappa_2, \ldots, \varkappa_h$ be h primals on $V_d^{(n)}$, where $2 \leq h \leq d - 1$, such that they meet in a variety μ, with regular dimension $d - h$, which consists of various components $\mu_1, \mu_2, \ldots, \mu_l$ having respectively orders $m_1 n^{d-h}, m_2 n^{d-h}, \ldots, m_l n^{d-h}$ and multiplicities $p_1, p_2, \ldots, p_l$ in μ. With the notation of the final proposition of § 57, p. 124, we obtain then the relation

$$m_1 p_1 + m_2 p_2 + \cdots + m_l p_l = k_1 k_2 \ldots k_h.$$

Let $K_1, K_2, \ldots, K_h$ be the hypersurfaces of S_d which are the images of $\varkappa_1, \varkappa_2, \ldots, \varkappa_h$, and let $r - t - 1$ be the dimension of the linear system

$(K_1,K_2,\ldots,K_h)_n$. Then, with the varieties $\varkappa_1,\varkappa_2,\ldots,\varkappa_h$ there is associated a unique space, S_t, which is the image in S_r of this linear system (§ 56, p. 122). We wish to construct S_t by suitable considerations with respect to the varieties $\varkappa_1,\varkappa_2,\ldots,\varkappa_h$.

In view of the theorem of Severi (§ 55, p. 119), it is evident that, *if the variety μ is free from multiple components*, i. e. each of the p_i's is equal to 1, *S_t is none other than the ambient of μ*. This is no longer true when at least one of the p_i's exceeds unity. We shall see that, in this case, we can intrinsically associate with each component μ_i of μ a variety, μ_i^*, of S_t, which passes through μ_i, such that S_t is precisely the ambient space of the variety

$$\mu^* = \mu_1^* + \mu_2^* + \cdots + \mu_l^* .$$

In order to define such a variety μ_i^*, we fix in S_d a space S_{h-1}, generally situated with respect to $K_1,K_2,\ldots,K_h$. Hence, by the final proposition of § 55, p. 121, in order that a hypersurface of S_d should belong to the module $(K_1,K_2,\ldots,K_h)$, it is necessary and sufficient that its section by a generic S_h passing through S_{h-1} should belong to the module determined by the sections of $K_1,K_2,\ldots,K_h$ with S_h. This condition can be interpreted in S_r, when we think of S_h as the intersection of $d-h$ generic hyperplanes of S_d passing through S_{h-1}, and then apply the results of § 58, p. 126 to the d hypersurfaces of S_r consisting of $K_1,K_2,\ldots,K_h$ and the above $d-h$ hyperplanes. We thus obtain, without difficulty, the following theorem.

With the hypotheses of the second paragraph of the present section, let $\sum_{d-h}$ be a fixed generic linear subsystem, with dimension $d-h$, of the complete linear system $\sum_d$ (with dimension d) which comprises the primals on $V_d^{(n)}$ with minimal order ($= n^{d-1}$). If $\varkappa_{h+1},\varkappa_{h+2},\ldots,\varkappa_d$ are $d-h$ general primals of $\sum_{d-h}$, then the primals $\varkappa_1,\varkappa_2,\ldots,\varkappa_h,\varkappa_{h+1},\ldots,\varkappa_d$ intersect in $k_1k_2\ldots k_h$ points, divided into l sets of points lying on the various components of μ. More precisely, on the component μ_i the set consists of m_i points, each counted with multiplicity p_i in the above intersection ($i = 1, 2, \ldots, l$). In virtue of § 58, p. 126, there is, intrinsically associated with such a point, an S_{p_i-1} which only depends on the given primals $\varkappa_1,\varkappa_2,\ldots,\varkappa_h$ of $V_d^{(n)}$ and on the ∞^{d-h-1} linear system determined by $\varkappa_{h+1},\varkappa_{h+2},\ldots,\varkappa_d$. The spaces S_{p_i-1} generate an algebraic variety, μ_i^*, as this ∞^{d-h-1} linear system varies in $\sum_{d-h}$: μ_i^* in general has dimension $d-h+p_i-1$. The space S_t containing μ, which is defined by the module $(K_1,K_2,\ldots,K_h)$, is precisely the minimal linear space which contains each of the varieties μ_i^* ($i = 1, 2, \ldots, l$).

We note that there remains the question of the algebro-differential relations which connect the varieties μ_i^*, μ_i and $V_d^{(n)}$, and also that it might be of interest to investigate the algebraic system described by the μ_i^*'s as $\sum_{d-h}$ varies in $\sum_d$.

§ 60. A special property of the space associated with an isolated intersection on $V_d^{(n)}$ in the simple case. — We return to the situation described in § 58, p. 125, so that $\varkappa_1, \varkappa_2, \ldots, \varkappa_d$ are d primals on $V_d^{(n)}$ having an isolated intersection at O_i. If O_i is counted in the intersection of these primals with multiplicity p_i, then, when we have some precise knowledge of the behaviour of $\varkappa_1, \varkappa_2, \ldots, \varkappa_d$ at O_i, we can often deduce more definite results regarding the position of the associated S_{p_i-1} than those enunciated in § 58, p. 126. Such considerations are of a local character, so that the suffix i can be conveniently omitted. We can for example prove the following result:

With the notations and hypotheses of § 58, *p.* 125, *we suppose also that* $d \geqq 2$ *and that the primals* $\varkappa_1, \varkappa_1, \ldots, \varkappa_d$ *present the simple case at a common point, O. In other words, if* $t_1, t_2, \ldots, t_d$ $(\geqq 1)$ *are the multiplicities of* $\varkappa_1, \varkappa_2, \ldots, \varkappa_d$ *respectively, at O, then the multiplicity of intersection of these varieties at O is supposed to be* $p = t_1 t_2 \ldots t_d$. *If also*

$$n \geqq t_1 t_2 \ldots t_d + (t_1 + t_2 + \cdots + t_{d-1}) - d\,, \tag{31}$$

then the space S_{p-1}, *associated with this intersection at O, necessarily belongs to the* $(t_1 + t_2 + \cdots + t_d - d)$*-osculating space to* $V_d^{(n)}$ *at O.*

It will suffice to prove this result on the additional hypothesis that the tangent cones to $\varkappa_1, \varkappa_2, \ldots, \varkappa_{d-1}$ at O meet in $g = t_1 t_2 \ldots t_{d-1}$ distinct generating lines, $l_1, l_2, \ldots, l_g$. Indeed, we can pass from this simple case to that in which the generators have any manner of coincidence, by means of suitable limiting considerations with respect to S_{p-1} and the $(t_1 + t_2 + \cdots + t_d - d)$-osculating space.

We shall suppose, then, that the curve of intersection of $\varkappa_1, \varkappa_2, \ldots, \varkappa_{d-1}$ has g linear branches $L_1, L_2, \ldots, L_g$ at O, and that the branch tangents are the g lines l_i. Each of these branches corresponds to a linear branch in S_d, whence we deduce that it is n-regular at O. We see from § 58, p. 125, that, in order to construct S_{p-1}, we can proceed as follows. We displace $\varkappa_d$ generically into a neighbouring variety, $\bar{\varkappa}_d$, and consider the g sets of t_d distinct points cut by $\bar{\varkappa}_d$ on the linear branches L_i. We thus obtain $g\, t_d = t_1 t_2 \ldots t_{d-1} t_d = p$ mutually independent points all different from O, and these points are therefore joined by a space $\bar{S}_{p-1}$. As $\bar{\varkappa}_d$ tends to $\varkappa_d$, this space has a well-defined limit, which is the required space S_{p-1}.

We abbreviate by writing

$$a = t_1 + t_2 + \cdots + t_{d-1} - d, \quad b = d_a - 1\,,$$

and we refer to the a-osculating space to $V_d^{(n)}$ at O as S_b. In virtue of the second theorem of § 56, p. 122, and using the inequality (31), we deduce that S_b and $\bar{S}_{p-1}$ are independent; they are therefore joined by a space $\bar{S}_{b+p}$. Let S_{b+p} be any position of accumulation of $\bar{S}_{b+p}$ as $\bar{\varkappa}_d \to \varkappa_d$.

Now, by construction, $\bar{S}_{b+p}$ contains the space of dimension a which osculates the branch L_i at O $(i = 1, 2, \ldots, g)$, and, moreover, it also contains the t_d points of this branch which tend to O. Therefore (§ 51, p. 110), S_{b+p} must contain the space, $S^{(i)}$, of dimension $a + t_d$ which osculates L_i at O. We recall (§ 54, p. 118, (21), (23)) that, for the star of lines with centre O in the tangent S_d to $V_d^{(n)}$ at O, the g lines l_i impose distinct conditions on the cones with vertex O and order $\geqq a+1$ which must contain them.

We shall now prove that the space, U, which joins $S_b, S^{(1)}, S^{(2)}, \ldots, S^{(g)}$ has dimension

$$b + g\, t_d = b + p\,.$$

We shall proceed by induction with respect to t_d. Thus, firstly, we note that the result is certainly true when $t_d = 0$, since in this case each of the spaces $S^{(i)}$ lies in S_b. We thus assume that $t_d \geqq 1$ and that the space, V, joining S_b to the spaces $T^{(i)}$ of dimension $a + t_d - 1$, which osculate the L_i's at O $(i = 1, 2, \ldots, g)$, has dimension

$$b + g\,(t_d - 1) = b + p - g\,.$$

Hence the space U, considered as the space which joins V to the g spaces $S^{(i)}$, has dimension

$$(b + p - g) + g = b + p\,,$$

unless one of the $S^{(i)}$'s is contained in the space which joins V to the remaining $S^{(i)}$'s. In order to obtain a contradiction, we thus suppose, for example, that $S^{(1)}$ lies in the space $V S^{(2)} \ldots S^{(g)} = S_b T^{(1)} S^{(2)} \ldots S^{(g)}$. We shall show that this cannot in fact occur, by proving the existence of a hyperplane of $S_r = S_{d_n - 1}$ which contains the second but not the first of these spaces. We obtain such a hyperplane as the image, in S_{d_n-1}, of a form of order n in S_d which at O' — the homologue of O — has multiplicity $a + t_d$ $(\geqq a + 1)$, such that the associated tangent cone at O' passes through the directions at O' which correspond to $l_2, l_3, \ldots, l_g$, without passing through that corresponding to l_1. Such a hyperplane certainly exists, in view of the hypotheses concerning n, and the remark at the end of the preceding paragraph.

Hence the space $U = S_b S^{(1)} S^{(2)} \ldots S^{(g)}$ has dimension $b + p$. But, from the preceding observations, U must lie in S_{b+p}: therefore U evidently coincides with S_{b+p}. Further $U = S_{b+p}$ manifestly lies in the $(a + t_d)$-osculating space to $V_d^{(n)}$ at O, since this property is enjoyed by each $S^{(i)}$ and by S_b. The required result is now immediate, when we note that

$$a + t_d = t_1 + t_2 + \cdots + t_d - d\,,$$

and that S_{b+p} must pass through S_{p-1}, since $\bar{S}_{b+p}$ contain $\bar{S}_{p-1}$ by construction.

§ 61. On a theorem of Torelli and some complements. — We shall first prove the following theorem.

Let $K_1, K_2, \ldots, K_h$ be h hypersurfaces of S_d $(2 \leq h \leq d)$ which intersect in a variety M of regular dimension $d - h$, where M has components of any multiplicity and $K_1, K_2, \ldots, K_h$ behave arbitrarily along each of these components. If a hypersurface, N, of S_d passes through each component of M with multiplicity not less than the multiplicity of that component in M, then N necessarily belongs to the module $(K_1, K_2, \ldots, K_h)$.

We preface the proof of this result with two observations. Firstly, if $h < d$, we denote by S'_h a generic space of dimension h in S_d, and by $K'_1, K'_2, \ldots, K'_h, M', N'$ the sections of $K_1, K_2, \ldots, K_h, M, N$ respectively by the space S'_h. The hypotheses of the theorem now hold for the primed symbols, if we put $d' = h$. Hence, if the theorem is assumed under this condition, we see that N' belongs to $(K'_1, K'_2, \ldots, K'_h)$. Therefore it now follows, from the penultimate proposition of § 55, p. 120, that N must in fact belong to the module $(K_1, K_2, \ldots, K_h)$.

Secondly, if the result has been proved for forms N of degree n, where n is sufficiently high (with respect to the projective characters of $K_1, K_2, \ldots, K_h$), for $n \geq n_0$ say, then the theorem itself follows immediately for any form of degree $n < n_0$. For it suffices to consider a generic form A of degree $n_0 - n$, and to apply the partial result to the form AN, which has degree n_0. We then deduce that AN belongs to $(K_1, K_2, \ldots, K_h)$; so that N also belongs to this module, in virtue of the second theorem of LASKER (§ 54, p. 118).

In view of the above remarks, it will suffice to prove the theorem for the case in which $h = d$, and the inequality (30) holds. On these hypotheses, the observations of § 58, p. 126, apply to the present situation, and we can avail ourselves of the theorem proved in that section. To the form N, there corresponds in S_r the section of $V_d^{(n)}$ by a certain S_{r-1}. Now, *to say that N belongs to $(K_1, K_2, \ldots, K_d)$, is equivalent to saying that S_{r-1} passes through the space S_{m-1} (in the notation of § 58, p. 125), i. e., that S_{r-1} contains each of the spaces S_{p_i-1}* $(i = 1, 2, \ldots, s)$. Certainly this last condition is satisfied in the present case. Indeed, recalling the observations of § 56, p. 122, we see that S_{r-1} contains the $(p_i - 1)$-osculating space to $V_d^{(n)}$ at the point O_i (for $i = 1, 2, \ldots, s$); and this last space contains S_{p_i-1}, in view of the theorem of § 58, p. 126. Thus the assertion is proved.

It is now a simple matter to establish the theorem of TORELLI stated in § 55, p. 120. The proof of this theorem is completely analogous to that given for the preceding theorem. Indeed, only the final paragraph of the above proof need be changed; more precisely, we would use the results of § 60, p. 128 instead of those of § 58, p. 126, and employ the relation (31) in place of (30). In fact, in both of these proofs it would suffice to assume $n \geqq k_1 k_2 \ldots k_d + (k_1 + k_2 + \cdots + k_d) - d$, in which case (31) and (30) are both satisfied.

We note that, in view of the local character of the methods used for the proofs of these two theorems, we can immediately combine the two results, in order to obtain a theorem relative to the *moduli defined by forms which intersect regularly, and present the simple case in only some of the components of their variety of intersection.*

We close this part with some remarks concerning the local conditions which must be imposed on a form in order that it should belong to a given module. *With the hypothesis* $h = d$, and using the above terminology, we say that the local necessary and sufficient conditions for N to belong to $(K_1, K_2, \ldots, K_d)$ are that the corresponding S_{r-1} should contain the spaces S_{p_i-1} $(i = 1, 2, \ldots, s)$. However, in general, these spaces admit of no simple definition relative to O_i and $V_d^{(n)}$ only; we must therefore be content with sufficient conditions, where the S_{p_i-1}'s are replaced by suitable spaces containing them, which have a certain behaviour at O_i. Such behaviour is naturally dependent on the manner in which the varieties $\varkappa_1, \varkappa_2, \ldots, \varkappa_d$ intersect at O_i. In this way we could obtain results other than those enunciated in §§ 58, 60, pp. 126, 128. Finally, *with the hypothesis* $h < d$, and employing the notation and hypotheses of § 59, we see that the foregoing applies similarly relative to the ambient spaces of the varieties μ^*.

Historical Notes and Bibliography

The present Part is, apart from some minor modifications, a translation of B. SEGRE [120].

The varieties $\{h, k\}$ associated with an n-regular point of a d-dimensional variety, where $0 \leqq h < k \leqq n$, which here appear in § 50, p. 109, were introduced by DEL PEZZO [42] and C. SEGRE [127], for the smallest values of d, h, k. The methods of these two authors were respectively of a synthetic and an analytic character. Many authors have investigated the properties of these varieties since their introduction: cf. particularly C. SEGRE [128, 129], SEVERI [143], BOMPIANI [17], LEVI-SANTALÓ-DE MARIA [70], and LONGO [72]. The final result of § 50, p. 110, here obtained from the study of the above varieties, has been given in the most elementary case $d = 2$, $h = 1$ in BOMPIANI [19] by means of a direct calculation.

The general procedure outlined in § 50 can be also used in the consideration of the regular or irregular differential elements, of dimension less than d, on the differentiable variety W_d; for the case in which W_d is a plane, see LONGO [73] and the works there cited.

In §§ 52—55, pp. 114—121, it has been found convenient to give references, relevant to the various well-known algebraic theorems (appearing in the literature before B. SEGRE [120]), as such theorems are enunciated.

Some differential properties of VERONESE varieties are known (cf. B. SEGRE [99, 110]), but a systematic investigation of such properties is still lacking. In the final six sections we begin such an investigation, commencing with the algebraic results of §§ 52—55. The properties we have derived have a local rather than a global character, and could, as we observed in § 50, p. 109, be used in the study of topological invariants relative to the subvarieties of a variety, even in the non-algebraic cases.

Our considerations of the VERONESE varieties offer geometric illustrations of classical results concerning the modules generated by algebraic forms, in particular, for the theorem of SEVERI [131] and the theorem of TORELLI [160] — which extends those of NÖTHER and KÖNIG — illustrated in § 57, p. 124 and § 61, p. 130 respectively.

In this respect, we give here a new interpretation and many extensions of the recent results of TIBILETTI [159]. Fundamental to such considerations is the theorem of § 51, p. 100; this theorem completely justifies the theor. 4 of [159], and that was necessary, since the proof which accompanies the latter seems to be inconclusive. In § 61, p. 161, among other things, we give a more precise meaning to the above author's concept of a form locally belonging to a module, and at the same time generalize the notion to any dimension.

Part Eight

Linear Partial Differential Equations

Although we shall mainly be concerned in this Part with *differential equations*, the methods we use here for their discussion and solution are intimately connected with the geometry of the rest of the volume. In particular, the results obtained depend to a great extent on the theory of modules and the intersections of a set of algebraic varieties, considered in Part Seven and, occasionally, in Parts Five and Six. This arises essentially from the fact that there exists an isomorphism between the commutative ring of linear differential operators with constant coefficients and the ring of polynominals in a finite number of indeterminates.

§ 62. Preliminary observations. — In §§ 63—65, we shall conduct all our considerations in the real number field; however, it should be noted that the results obtained could be easily extended to other commutative fields, for example the field of complex numbers. The symbols y, z, and u, v (sometimes with the addition of suffixes) will be used to denote functions of a single variable x, the first two being unknown functions and the latter being known. Such functions we shall assume to be differentiable as many times as is necessary in order that all the expressions which arise should have a meaning.

A differential operator of the type

$$\alpha = \alpha\left(\frac{d}{dx}\right) = a_0 \frac{d^k}{dx^k} + a_1 \frac{d^{k-1}}{dx^{k-1}} + \cdots + a_{k-1} \frac{d}{dx} + a_k ,$$

where the a_i's are constants (real numbers), will be called briefly *a differential polynominal*. This operator is of order k, when $a_0 \neq 0$, and in this case a_0 is said to be its *leading coefficient*.

Hence a linear differential equation with constant coefficients:

$$a_0 \frac{d^k y}{dx^k} + a_1 \frac{d^{k-1} y}{dx^{k-1}} + \cdots + a_{k-1} \frac{dy}{dx} + a_k y = u , \tag{1}$$

can be written simply as

$$\alpha y = u .$$

It is well known that, if $a_0 \neq 0$, the general solution of (1) depends on k arbitrary constants, and it can be obtained by integration preceded by the resolution of the polynomial equation $\alpha(x) = 0$, of order k (the so-called characteristic equation of (1)).

We shall first consider systems of n ($\geqq 1$) ordinary linear differential equations, with constant coefficients, in m ($\geqq 1$) unknowns. By an obvious extension of the previous notation, we can write the general system of this type in the form

$$\sum_{1}^{m}{}_j \alpha_{ij} y_j = u_i \qquad (i = 1, 2, \ldots, n). \quad (2)$$

This can be represented more compactly by the single matrix equation

$$A\,Y = U, \qquad (3)$$

where A is the $n \times m$ matrix whose elements are the differential polynomials α_{ij}, and Y, U are the column vectors formed by the functions $(y_1, y_2, \ldots, y_m)$ and $(u_1, u_2, \ldots, u_n)$.

§ 63. The reduction of differential equations to a canonical form. — Let N be a square matrix of order n whose elements, $\nu_{hi} = \nu_{hi}\left(\frac{d}{dx}\right)$ $(h, i = 1, 2, \ldots, n)$, are arbitrary differential polynomials. The system (2) thus gives rise to the system represented by

$$NAY = NU; \qquad (4)$$

indeed, the h-th equation of this system we obtain by operating on each side of the i-th equation of (2) with ν_{hi} and then summing over the index $i = 1, 2, \ldots, n$.

We note that the determinant of the matrix N is always a differential polynomial. If this polynomial is in fact a *non-zero constant*, then there exists also an inverse matrix N^{-1} whose elements are differential polynomials. With this hypothesis, the systems (3) and (4) are equivalent, since (3) is obtained from (4) when we multiply each side of the latter on the left by N^{-1}. By a known algebraic result (cf. for example SCHREIER-SPERNER [89], §8) we can, with the above hypothesis, obtain (4) from (3) by suitably using a finite number of times the following two elementary operations:

a) any equation of the system is replaced by the equation obtained on multiplying it by a non-zero constant;

b) any equation of the system is replaced by the equation obtained on adding to it the result of an arbitrary differential polynomial applied to any other equation of the system.

We note further that the problem of solving (3) is in essence unaltered, when the unknowns $(y_1, y_2, \ldots, y_m)$ are replaced by other unknowns $(z_1, z_2, \ldots, z_m)$ which are connected with them by

$$Y = MZ, \qquad (5)$$

where M is a square matrix of order m, whose elements are differential polynominals, such that the determinant of M is a *non-zero constant.* Indeed, in this case equations (5) can be solved uniquely for the z_i's, giving

$$Z = M^{-1} Y$$

From the algebraic result used above, we can pass from the old unknowns to the new ones (and conversely), by suitably applying the following two elementary operations a finite number of times:

a′) y_i is replaced by $c y_i$, where c is any non-zero constant;

b′) y_i is replaced by $y_i + \gamma y_j$, where $j \neq i$, and γ denotes any differential polynomial.

We now apply (5) to the system (4), and so obtain a system

$$BZ = V, \tag{6}$$

which is equivalent to (3), where for brevity we have put

$$B = NAM, \tag{7}$$

$$V = NU. \tag{8}$$

Making use of another algebraic property concerning matrices with polynomial elements, we see that in (7) N and M can be chosen so that B takes the canonical form in which the only non-zero elements are the first r elements of the principal diagonal, where r is the rank of A. Moreover, M and N can be chosen so that these elements are $\beta_1, \beta_2, \ldots, \beta_r$, the so-called *elementary divisors* of A, which are (non-zero) differential polynomials with unit leading coefficients satisfying

$$\beta_1 \mid \beta_2 \mid \ldots \mid \beta_r, \tag{9}$$

where here, and in the sequel, we put $\varrho \mid \sigma$ to signify that ϱ and σ are two polynomials such that ϱ is a factor of σ. The algebraic property used in this reduction is a simple generalization to rectangular matrices of a result which is well-known (cf., for example, SCHREIER-SPERNER [89], § 8, or WEDDERBURN [169], pp. 33—37) for the case in which A is square.

The calculation of the polynomials $\beta_1, \ldots, \beta_r$ can be effected in the following manner. Let δ_k be the differential polynomial, with unit leading term, which is the H. C. F. of the minors of order k extracted from A ($k = 1, 2, \ldots, r$). Put also $\delta_0 = 1$, so that we have

$$\delta_0 \mid \delta_1 \mid \ldots \mid \delta_r;$$

β_k is then given by

$$\beta_k = \frac{\delta_k}{\delta_{k-1}} \qquad (k = 1, 2, \ldots, r). \tag{10}$$

§ 64. Remarks on the solution of the differential equations. — When B is reduced to its canonical form as above, the system of equations represented by (6) obviously divides into two sets

$$\beta_k z_k = v_k \qquad (k = 1, 2, \ldots, r), \quad (11)$$

$$v_l = 0 \qquad (l = r + 1, r + 2, \ldots, n), \quad (12)$$

the second of which is lacking if, and only if, $r = n$. Recalling the relations (8), we see that (12) gives a set of $n - r$ distinct linear homogeneous differential equations with constant coefficients which must be satisfied by the known functions $u_1, u_2, \ldots, u_n$, which appear as the right-hand sides of (2), if the equations (2) are to possess a solution. Thus we can call these relations the conditions of integrability of the given system (2). When these relations are satisfied, it suffices to consider only the r differential equations (11) in order to find the solutions of (2). In the k-th equation of (11), z_k is the only unknown which enters $(k = 1, 2, \ldots, r)$: hence, if $m > r$, the unknowns $z_{r+1}, z_{r+2}, \ldots, z_m$ can be chosen arbitrarily.

As has been observed earlier, the separate equations of (11) can be solved by means of integration and the extraction of the roots of certain polynomial equations. The solution of these equations is accompanied — in a well-known manner — by the intervention of a number of arbitrary essential constants; this number is equal to the sum of the degrees of the polynomial β, i. e., from (10), the degree of the polynomial

$$\beta_1 \beta_2 \ldots \beta_r = \delta_r .$$

Moreover, the integration of those equations is obtained from the resolution of the algebraic equations

$$\beta_1(x) = 0, \quad \beta_2(x) = 0, \quad \ldots, \quad \beta_r(x) = 0 ,$$

the roots of which, from (9), are always roots of

$$\beta_r(x) = 0 . \quad (13)$$

We note further that the degree of β_r is equal to the difference of the degrees of δ_r and δ_{r-1}, in view of the final equation of (10).

We have thus obtained the following result, part of which has previously been established, with another proof, by Picone and Ghizzetti [84].

Given a system, (2), of n ordinary linear differential equations with constant coefficients in m unknown functions $y_1, y_2, \ldots, y_m$, where the α_{ij}'s are mn arbitrary differential polynomials, and the $u_1, u_2, \ldots, u_n$ are assigned functions, let r be the rank of the matrix A with elements α_{ij}. The equations are compatible if, and only if, the u_i's satisfy a certain set of $n - r$ distinct linear homogeneous differential equations with constant coefficients (these conditions, of course, no longer appear if $n = r$). When these conditions are satisfied, the general integral of (2) can be obtained by algebraic operations and integration, and depends on $m - r$

arbitrary functions and on a set of arbitrary constants equal in number to the degree of δ_r, *the H. C. F. of the minors of order* r *extracted from* A. *The above algebraic operations reduce essentially to the resolution of a single algebraic equation,* (13), *the degree of which is equal to the difference of the degree of* δ_r *and of* δ_{r-1}, *the analogous polynomial for the minors of order* $r-1$ *(taken equal to unity if* $r = 1$).

§ 65. The construction of the conditions of integrability. — The procedure outlined in the preceding two sections gives a practical method for the formation of the conditions of integrability (12), and also for the integration of the given system (2). This procedure relies on the determination of the two matrices M, N, by means of which A is reduced to its canonical form. In virtue of § 63, p. 133, this is equivalent to obtaining (11), (12) from (2) by means of a suitable finite sequence of operations a), b), and a'), b'). This process can, however, be rather lengthy; consequently, it is convenient to suggest a method by which (12) can be obtained more quickly.

We introduce two row vectors $R = (\varrho_1, \varrho_2, \ldots, \varrho_n)$ and $S = (\sigma_1, \sigma_2, \ldots, \sigma_n)$, where the ϱ_i's and σ_i's are differential polynomials which we suppose for the moment to be connected only by

$$S = RN,$$

or

$$R = SN^{-1}. \tag{14}$$

From (6) we see that:

$$RV = 0, \tag{15}$$

for *every* vector R which satisfies

$$RB = 0. \tag{16}$$

Moreover, (15) is *equivalent* to

$$SU = 0, \tag{17}$$

from (14) and (8); further (16) is equivalent to

$$SA = 0, \tag{18}$$

from (14) and (7).

On the other hand, since B is in the canonical form, (16) is satisfied if, and only if,

$$\varrho_1 = \varrho_2 = \cdots = \varrho_r = 0.$$

On this hypothesis, (15) reduces to

$$\varrho_{r+1} v_{r+1} + \varrho_{r+2} v_{r+2} + \cdots + \varrho_n v_n = 0;$$

and this is manifestly equivalent to (12), since $\varrho_{r+1}, \varrho_{r+2}, \ldots, \varrho_n$ are arbitrary: relations (12) are in fact obtained by putting, in turn, one of the ϱ_i's $(r+1 \leq i \leq n)$ equal to unity and the others zero. Thus we obtain conditions equivalent to (12) when we state that (17) must hold for every vector S satisfing (18). Hence:

A necessary and sufficient condition that the system (2), *with the rank r of A less than n, should possess a solution, is that the u_i's should satisfy*

$$\sigma_1 u_1 + \sigma_2 u_2 + \cdots + \sigma_n u_n = 0\,, \tag{19}$$

whenever the differential polynomials $\sigma_1, \sigma_2, \ldots, \sigma_n$ are chosen such that identically

$$\sum_{1}^{n}{}_i\, \alpha_{ij}\sigma_i = 0 \qquad (j = 1, 2, \ldots, m)\,. \tag{20}$$

For this, is suffices to consider (19) *for each of a set of $n-r$ choices of the σ_i's, when such a set forms a primitive system (which always exists) of solutions of* (20) *(i. e., so that any other solution of* (2) *can be obtained as a linear combination with polynomial coefficients of the above $n-r$ solutions).*

§ 66. The conditions of compatibility for a system of linear partial differential equations in one unknown. — We shall now consider differential equations in the d variables $x_1, x_2, \ldots, x_d$. Hence u, v and w (sometimes with distinguishing suffixes affixed) will at present denote *known* functions of $x_1, x_2, \ldots, x_d$, which are defined in a connected region of the space S_d for which the x_i's are current coordinates. These functions will again be assumed to be differentiable as many times as is necessary; y will denote similarly an unknown function of x. We shall put briefly

$$y^{k_1 k_2 \cdots k_d} = \frac{\partial^{k_1+k_2+\cdots+k_d}\, y}{\partial x_1^{k_1}\, \partial x_2^{k_2} \cdots \partial x_d^{k_d}}\,,$$

and, in particular, $y^{00\cdots 0} = y$.

As a simple extension of the notation of § 62, p. 132, we shall call a differential operator of the type

$$\alpha = \sum_k a_{k_1 k_2 \cdots k_d} \frac{\partial^{k_1+k_2+\cdots+k_d}}{\partial x_1^{k_1}\, \partial x_2^{k_2} \cdots \partial x_d^{k_d}}\,,$$

where the a_k's are constants, a differential polynomial, when the summation ranges over a finite number of terms. There is clearly an isomorphism between the commutative ring of such polynomials and that of the ordinary polynomials in d variables:

$$\mathfrak{a} = \mathfrak{a}(x_1, x_2, \ldots, x_d) = \sum_k a_{k_1 k_2 \cdots k_d} x_1^{k_1}\, x_2^{k_2} \ldots x_d^{k_d}\,.$$

If such a polynomial does not reduce to its constant term, we shall also denote by $\mathfrak{a}$ the hypersurface of S_d which has equation

$$\mathfrak{a}(x_1, x_2, \ldots, x_d) = 0\,.$$

We shall say that $\mathfrak{a}$ and α are associated, and that $\mathfrak{a}$ is the characteristic hypersurface of the operator α.

With the above notation, the most general linear partial differential equation with constant coefficients is given simply by $\alpha y = u$; this is obviously an abbreviated form of the equation

$$\sum_k a_{k_1 k_2 \cdots k_d} y^{k_1 k_2 \cdots k_d} = u .$$

We propose first to study the conditions of compatibility for a finite system of h ($\geqq 2$) equations of this type, each of which is an equation in the same unknown, y.

For the system

$$\alpha_1 y = u_1, \quad \alpha_2 y = u_2, \ldots, \quad \alpha_h y = u_h , \tag{21}$$

we immediately obtain a set of $h(h-1)/2$ such *necessary* conditions, namely:

$$\alpha_i u_j = \alpha_j u_i \qquad (i, j = 1,2, \ldots, h) . \tag{22}$$

We shall discuss if, and when, these conditions are *sufficient.* It will be seen in the sequel that the answer to this question is essentially dependent on certain properties of the module $(\mathfrak{a}_1, \mathfrak{a}_2, \ldots, \mathfrak{a}_h)$ which is generated by the polynomials, $\mathfrak{a}_i$, associated with the operators α_i appearing in (21). It will be found convenient to write (22) in the form

$$v_{ij} = v_{ji} \qquad (i, j = 1, 2, \ldots, h), \tag{23}$$

where we define

$$v_{ij} = \alpha_j u_i . \tag{24}$$

A very simple case in which the relations (22) suffice for the integrability of the system (21), is that in which *the algebraic hypersurfaces* $\mathfrak{a}_1, \mathfrak{a}_2, \ldots, \mathfrak{a}_h$ *of* S_d *have no point (either real or complex) in common in the finite part of the space.* With this hypothesis, and if (22) hold, the system of differential equations (21) have one, and only one, solution. Further we can calculate this solution, as follows, by means of operations involving only differentiation (see B. SEGRE [106], n. 1). Since the varieties $\mathfrak{a}_1, \mathfrak{a}_2, \ldots, \mathfrak{a}_h$ have no finite point in common, there exist differential polynominals, λ, such that

$$\lambda_1 \alpha_1 + \lambda_2 \alpha_2 + \cdots + \lambda_h \alpha_h = 1$$

(cf. PERRIN [82], and also e. g. VAN DER WAERDEN [165], p. 5). Hence the above solution is given by

$$y = \lambda_1 u_1 + \lambda_2 u_2 + \cdots + \lambda_h u_h .$$

It is however possible to construct as many examples as we wish such that (21) *have no solution,* although the relations (22) hold. For this, it is sufficient to observe that any *syzygy*

$$\mathfrak{l}_1 \mathfrak{a}_1 + \mathfrak{l}_2 \mathfrak{a}_2 + \cdots + \mathfrak{l}_h \mathfrak{a}_h = 0$$

between the polynomials $\mathfrak{a}_i$ gives rise to an identity between the associated differential polynomials, of the form

$$\lambda_1 \alpha_1 + \lambda_2 \alpha_2 + \cdots + \lambda_h \alpha_h = 0 . \tag{25}$$

Whence, in order that the system (21) should have a solution, it is necessary that

$$\lambda_1 u_1 + \lambda_2 u_2 + \cdots + \lambda_h u_h = 0: \tag{25'}$$

but this differential relation between the known functions u_i is not always a consequence of (22). This can be immediately seen from the case in which the λ's are non-zero constants, and also when $\alpha_1 = \beta_1 \gamma$ and $\alpha_2 = \beta_2 \gamma$ (where γ is a differential polynomial not reduced to a constant), so that now, on assuming $\lambda_1 = \beta_2, \lambda_2 = -\beta_1$ and $\lambda_3 = \cdots = \lambda_h = 0$, (25) holds and (25′) is not a consequence of (22). Other examples will appear from § 70, p. 148.

In the following two sections, we shall establish further criteria of a quite general character for the sufficiency of the conditions of compatibility (22).

§ 67. The analytic case where the characteristic hypersurfaces intersect regularly. — We shall first prove the following theorem.

Relations (22) *are necessary and sufficient conditions for the existence of a solution of the partial differential equations* (21), *if:* (i) $u_1, u_2, \ldots, u_h$ *are analytic functions of the x's*; (ii) $2 \leq h \leq d$; (iii) *the characteristic algebraic hypersurfaces $\mathfrak{a}_1, \mathfrak{a}_2, \ldots, \mathfrak{a}_h$ of the operators $\alpha_1, \alpha_2, \ldots, \alpha_h$ have, in the complex d-dimensional projective space which contains them, an intersection, $\mathfrak{c}$, of regular dimension $d-h$.*

For the proof of this theorem, it is sufficient to show that the equations (21) are compatible with the hypotheses (i), (ii) and (iii), together with relations (23), where the v_{ij}'s are defined by (24). We note first that, from a known proposition of algebraic geometry (cf. p. 117, or Bertini [13], chap. XII, n. 16), it now follows that — if the α_i's satisfy an identity of the type (25) — there exist differential polynomials π_{ij} such that

$$\pi_{ij} = -\pi_{ji} \qquad (i, j = 1, 2, \ldots, h) \tag{26}$$

and

$$\lambda_i = \sum_{1}^{h}{}_j \, \pi_{ij} \alpha_j . \tag{27}$$

Further it is evident that, conversely, if (26) and (27) hold, the identity (25) follows.

On the other hand, from the condition (i) and from a classical property of systems of differential equations in the analytic field, we can obtain all the conditions of integrability for the system (21) by extending the system itself with successive arbitrary operations of differentiation (up to a certain order), and then eliminating the differentials $y^{k_1 k_2 \cdots k_d}$

amongst the equations so obtained. We notice now that the equations constructed in this manner contain the $y^{k_1 k_2 \cdots k_d}$'s linearly and with constant coefficients; consequently, the above elimination is carried out by linear combination of the equations with constant coefficients. Hence every condition of integrability is of the form

$$\lambda_1(\alpha_1 y - u_1) + \lambda_2(\alpha_2 y - u_2) + \cdots + \lambda_h(\alpha_h y - u_h) = 0,$$

or

$$(\lambda_1 \alpha_1 + \lambda_2 \alpha_2 + \cdots + \lambda_h \alpha_h) y = \lambda_1 u_1 + \lambda_2 u_2 + \cdots + \lambda_h u_h, \tag{28}$$

where the λ_i's are suitable differential polynomials. In fact these polynomials must satisfy a relation of the type (25), since it is necessary that no term $y^{k_1 k_2 \cdots k_d}$ should finally appear in (28).

Therefore, corresponding to any condition of integrability, there exist necessarily relations of the form (26), (27). Whence (28) reduces to

$$\lambda_1 u_1 + \lambda_2 u_2 + \cdots + \lambda_h u_h = 0 .$$

The theorem now follows, since this final relation is a consequence of (23). More precisely we have

$$\sum_1^h {}_i \lambda_i u_i = \sum_1^h {}_{ij} \pi_{ij} \alpha_j u_i = \sum_1^h {}_{ij} \pi_{ij} v_{ij},$$

and the final sum is identically zero, in virtue of (23) and (26).

The above result suggests further allied questions, which it may well be worth while to study. The development which follows clearly demonstrates the importance of the algebraic variety $\mathfrak{c}$, introduced in the preceding theorem, with regard to these problems.

We first distinguish between the two cases in which $\mathfrak{c}$ lies, or does not lie, completely in the hyperplane at infinity.

In the first case, we can apply the result given towards the end of § 66, p. 138, so that, when the conditions of integrability (22) are satisfied, *the system* (21) *admits precisely one solution, and this solution can be calculated by operations of differentiation only.*

In the second case, we can write $\mathfrak{c}$ in the form

$$\mathfrak{c} = \mathfrak{c}_f + \mathfrak{c}_i, \tag{29}$$

where $\mathfrak{c}_f$ is a pure non-empty $(d-h)$-dimensional algebraic variety (which may be reducible), of which no component lies completely at infinity, whilst $\mathfrak{c}_i$ (which may indeed be lacking) is the analogous variety lying completely at infinity. Keeping the notation and hypotheses of the preceding theorem, with the addition of the hypothesis that $\mathfrak{c}$ does not lie completely at infinity, we prove that:

If the conditions of integrability (22) *are satisfied, then the general integral of* (21) *depends on an infinite or a finite number of arbitrary constants according as $h < d$ or $h = d$. In the latter case, the number of* *arbitrary constants is equal to the number of points of the set* $\mathfrak{c}_f$, *each being counted with its multiplicity as a point of the intersection of the* $h = d$ *hypersurfaces* $\mathfrak{a}_1, \mathfrak{a}_2, \ldots, \mathfrak{a}_h$.

Let $(x_1, x_2, \ldots, x_d)$ be a point at which the u_i's are all holomorphic. We consider the TAYLOR expansion of a holomorphic solution, y, of the system (21) in a neighbourhood of this point. The coefficients $y^{k_1 k_2 \cdots k_d}$ of this expansion are uniquely subject to the condition of satisfying (21) and the equations we obtain by differentiating arbitrarily the elements of this system; and such an expansion represents then, in the neighbourhood of the point, the most general holomorphic solution of the given system.

We wish to determine how many of the $y^{k_1 k_2 \cdots k_d}$'s can be chosen freely, when they are subjected to the restrictions implied by the system (21) and equations deduced from this system by differentiation and linear combination with arbitrary constant coefficients. In view of the theorem proved above, the restrictions in which the $y^{k_1 k_2 \cdots k_d}$'s do not appear are all identically satisfied. From (28) we see that each of the remaining restrictions is of the form

$$\beta y = w\,, \tag{30}$$

where β is a differential polynomial (not identically zero) expressible in the form

$$\beta = \lambda_1 \alpha_1 + \lambda_2 \alpha_2 + \cdots + \lambda_h \alpha_h, \tag{31}$$

such that $\lambda_1, \lambda_2, \ldots, \lambda_h$ are differential polynomials, and

$$w = \lambda_1 u_1 + \lambda_2 u_2 + \cdots + \lambda_h u_h.$$

We remark that β cannot be a constant; for otherwise, by (31), the variety $\mathfrak{c}$ would lie at infinity. We can thus consider the hypersurface, $\mathfrak{b}$, associated with β. From (31), $\mathfrak{b}$ may be any variety arising from an element of the module $(\mathfrak{a}_1, \mathfrak{a}_2, \ldots, \mathfrak{a}_h)$, whence $\mathfrak{b}$ passes through each of the components of $\mathfrak{c}_f$.

Let n be an arbitrarily fixed non-negative integer. The number of distinct $y^{k_1 k_2 \cdots k_d}$'s such that $k_1 + k_2 + \cdots + k_d \leqq n$, is then precisely

$$d_n = \binom{n+d}{d}.$$

Let r_n be the maximum number of linearly independent equations (30) involving only the $y^{k_1 k_2 \cdots k_d}$'s where $k_1 + k_2 + \cdots + k_d \leqq n$. We note that precisely $d_n - r_n$ of the $y^{k_1 k_2 \cdots k_d}$'s, suitably chosen, can be given arbitrary values; and then the other differentials $y^{k_1 k_2 \cdots k_d}$, where $k_1 + k_2 + \cdots + k_d \leqq n$, are uniquely determined. From the geometric viewpoint, $d_n - 1$ is the dimension of the linear system of all the algebraic hypersurfaces $\mathfrak{b}$ of S_d with order n (the finite hypersurfaces with order $n' < n$ appearing in this system augmented by the hyperplane at infinity counted $n - n'$ times); similarly, $r_n - 1$ is the dimension of the linear system of hypersurfaces $\mathfrak{b}$ of the module $(\mathfrak{a}_1, \mathfrak{a}_2, \ldots, \mathfrak{a}_h)$ which have order n. Whence:

The number $d_n - r_n$ is the number of distinct conditions which must be imposed on a hypersurface $\mathfrak{b}$, of order n, in order that it should belong to $(\mathfrak{a}_1, \mathfrak{a}_2, \ldots, \mathfrak{a}_h)$.

Consequently, $d_n - r_n$ is not less than the number of conditions which must be imposed on $\mathfrak{b}$, in order that it should pass through the single components of the $(d-h)$-dimensional variety $\mathfrak{c}_f$. If $h < d$, this number, and so also $d_n - r_n$, increases indefinitely for increasing n. So that the first part of the theorem is proved.

We must now examine more closely the case $h = d$. Let $n_1, n_2, \ldots, n_d$ be the orders of the hypersurfaces $\mathfrak{a}_1, \mathfrak{a}_2, \ldots, \mathfrak{a}_d$ respectively; these are also the orders of the differential equations of the system (21). Suppose that

$$n \geqq n_1 n_2 \ldots n_d - 1 .$$

In virtue of an algebrico-geometric result of § 58, p. 125, the conditions mentioned in the preceding paragraph can be expressed as a system of conditions connected with the separate points of $\mathfrak{c}$, i. e., for the points of $\mathfrak{c}_f$ and the points of $\mathfrak{c}_i$. However, for the present case it is not necessary to take into account the conditions connected with the points at infinity, since they are automatically satisfied if, as is allowed, we adjoin to the hypersurface $\mathfrak{b}$ the hyperplane at infinity counted n times (§ 61, p. 130). This means that, in (31), the orders of β and λ_i are respectively not greater than n and $2n - n_i$; so that, as one can easily ascertain by examples, the order of each term on the right-hand side of (31) can be greater than the order of β. The single points of $\mathfrak{c}_f$, however, impose a set of conditions on $\mathfrak{b}$, of which the number of distinct conditions is equal to the multiplicity of the point in $\mathfrak{c}_f$ (see § 58, p. 126); this completes the proof of the theorem.

§ 68. An extension to the non-analytic case. — It would seem that many of the properties given in § 67, p. 139, can be extended to the non-analytic case. We shall, however, limit ourselves to showing that this is in fact the case when $h = d$, and this will lead to an extension of the preceding theorem.

We thus consider the system of differential equations

$$\alpha_1 y = u_1, \ \alpha_2 y = u_2, \ \ldots, \ \alpha_d y = u_d, \tag{32}$$

with orders $n_1, n_2, \ldots, n_d$ respectively, where for the u_i's we need only suppose that *the derivatives exist up to those of order $2n_1 n_2 \ldots n_d + 1$.* We suppose also that the conditions of integrability (22) (with $h = d$) are satisfied, and that the characteristic hypersurfaces $\mathfrak{a}_1, \mathfrak{a}_2, \ldots, \mathfrak{a}_d$ have only a finite number of points in common, of which precisely

m ($\leqq n_1 n_2 \ldots n_d$), say, lie in the finite part of the space. The number m is obtained, as usual, by counting each point with its proper multiplicity in the intersection of the $\mathfrak{a}_i$'s. We may suppose that m is positive, since $m = 0$ has already been dealt with fully in § 66, p. 138.

Putting $n = n_1 n_2 \ldots n_d - 1$, we see from § 67, p. 142 how, also with the present hypotheses, the complete system, $(30)_n$, with all the non-zero differential polynomials β of degree $\leqq n$ and of the type (31), may now be written; and, similarly, the analogous systems $(30)_{n+1}$ and $(30)_{n+2}$. By using the relations of $(30)_n$, we can express the various differentials $y^{k_1 k_2 \cdots k_d}$, with degree $k_1 + k_2 + \cdots + k_d \leqq n$, as functions of a suitable set of m such differentials. Denoting this set by $z_1, z_2, \ldots, z_m$, we see that, for all the values of the k_i's whose sum is not greater than n, $y^{k_1 k_2 \cdots k_d}$ is given by

$$y^{k_1 k_2 \cdots k_d} = f_{k_1 k_2 \cdots k_d}(z_1, z_2, \ldots, z_m) + w_{k_1 k_2 \cdots k_d}(x_1, x_2, \ldots, x_d), \tag{33}$$

where $f_{k_1 k_2 \cdots k_d}$ is a linear function of the z_i's with constant coefficients. We note that no equation of $(30)_n$ can contain $z_1, z_2, \ldots, z_m$ (or a subset of these), without containing some other $y^{k_1 k_2 \cdots k_d}$. It is also clear, from the definition of the z_i's, that amongst the equations of (33) there is precisely one for which the right-hand side consists simply of the term z_i ($i = 1, 2, \ldots, m$). The system $(30)_{n+1}$ evidently contains $(30)_n$ as a subsystem, and from $(30)_{n+1}$ we can express the various $y^{k_1 k_2 \cdots k_d}$ of degree $k_1 + k_2 + \cdots + k_d \leqq n + 1$ as functions of m of them. We remark that no equation of $(30)_{n+1}$ can contain some of $z_1, z_2, \ldots, z_m$ without containing another $y^{k_1 k_2 \cdots k_d}$, since such an equation would be an element of $(30)_n$, and we saw above that this cannot occur. Hence from $(30)_{n+1}$ we obtain the relations (33) and a set of similar relations for which $k_1 + k_2 + \cdots + k_d = n + 1$. Amongst this set we have equations of the form

$$\frac{\partial z_l}{\partial x_i} = f_{li}(z_1, z_2, \ldots, z_m) + w_{li}(x_1, x_2, \ldots, x_d) \qquad (l = 1, 2, \ldots, m;\ i = 1, 2, \ldots, d), \tag{34}$$

where the f_{li}'s are again linear functions of the z's with constant coefficients.

Differentiating (34) with respect to x_j, and eliminating the first derivatives of the z's by using (34), we obtain an expression

$$\frac{\partial^2 z_l}{\partial x_i \partial x_j} = f_{lij}(z_1, z_2, \ldots, z_m) + w_{lij}(x_1, x_2, \ldots, x_d),$$

which clearly belongs to $(30)_{n+2}$. A second expression for $\frac{\partial^2 z_l}{\partial x_i \partial x_j}$ can be obtained similarly from (34) by interchanging the roles of x_i and x_j. The right-hand sides of these two expressions must coincide identically; otherwise, we should either obtain a relation between the $z_1, z_2, \ldots, z_m$

and no other $y^{k_1 k_2 \cdots k_d}$, or else a relation between the variables x_i's, and neither of these possibilities can arise. This proves that the system (34) is *completely integrable.*

Therefore the unknowns z_i can be obtained, from equations (34), as functions of the x's by means of operations involving integrations only. In the general solution of this system there appear m arbitrary constants, given, for example, by the initial values of $z_1, z_2, \ldots, z_m$ (cf. MAYER [78] or e. g. SEVERI [136]). We now substitute the solutions just found into the equations (33), and so obtain the $y^{k_1 k_2 \cdots k_d}$'s, as functions of the x_i's and of the arbitrary constants, for all the values of the indices k such that the sum does not exceed n. By means of considerations analogous to those of the preceding paragraph, we deduce that, for these differentials,

$$y^{k_1 k_2 \cdots k_d} = \frac{\partial^{k_1+k_2+\cdots+k_d} y^{00\cdots 0}}{\partial x_1^{k_1} \partial x_2^{k_2} \cdots \partial x_d^{k_d}}.$$

Moreover, by construction, the expressions for the $y^{k_1 k_2 \cdots k_d}$'s given by (33) verify equations (32). Hence the function $y = y^{00\cdots 0}$ just obtained is the general solution of the system (32). We have now proved:

With the hypotheses stated in the second paragraph of this section (which are more general than those given in the enunciation of the theorem of § 67, *p.* 139 *when $h = d$), the general integral of the system* (32) *can be obtained by integration only, as indicated above. Such a solution depends on precisely m arbitrary constants, where m is the number of intersections of the characteristic hypersurfaces associated with the members of* (32) *which do not lie at infinity, each intersection being counted with its corresponding multiplicity.*

§ 69. Some remarks on sets of linear partial differential equations in several unknowns. — For the remainder of this Part, we shall consider the most general system of partial differential equations which are linear with constant coefficients. Such a system, of n equations in m unknown functions, is given by equations (2) of § 62, p. 133 and in the compact form of (3), where the α_{ij}'s now represent differential polynomials of the type indicated in § 66, p. 137. In the sequel, all the differential polynomials considered will be of this more general type, and we shall represent them by the Greek letters α, β, γ, δ, λ, μ, ν and ϱ.

We recall that in §§ 62, 63 we were considering a system of differential equations in a single variable, x, whereas we are now dealing with a similar system of partial differential equations in a set of d (≥ 2) variables, $x_1, x_2, \ldots, x_d$. Firstly we can apply, to this more general system, operations which closely resemble those applied in the simple case. Thus we can obtain from (3) — with its new significance — an equivalent system which can be written in the form (6), where A, U and Y are matrices analogous to those considered in §§ 62, p. 133, Z, B and V are given by (5), (7) and (8), and N, M are square matrices (of orders n and m) whose elements are differential polynomials and whose determinants are non-zero constants.

At this stage, the differences between the simple case of one variable and the present case of d ($\geqq 2$) variables become apparent. Indeed, it is now not possible, in general, to choose the matrices N, M such that B is reduced to the diagonal form; this occurs because it is not possible to extend the Euclidean division algorithm to polynomials in more than one variable. Moreover, it is no longer true that, for any choice of N, M, it is possible to pass from (3) to (6) by means of a finite sequence of elementary operations of the type a), b) and a'), b') of § 63, pp. 133—134. This observation suggests the algebraic problem of characterizing the matrices N, M which can be obtained, from the unit matrices of orders n and m, by means of operations on the rows and columns which are equivalent to the above elementary operations.

However, in practice, we can sometimes obtain a marked simplification of the given system by means of those elementary operations. Thus, by using the algebraic result of PERRIN previously employed in § 66, p. 138, we can establish:

A system of n linear partial differential equations with constant coefficients, in m unknowns and d variables, can generally be simplified if n and m exceed $d+1$. More precisely, if the constant coefficients of the n equations are general with respect to the system, then it is possible to construct an equivalent system of equations of $n-k$ equations in $m-k$ unknowns, where $k = \max(n-d-1, m-d-1)$.

We begin the proof of this theorem with the observation that, if some element α_{ij} of the matrix A is unity, then by means of a finite sequence of elementary operations, we can in effect remove all the other elements of the i-th row and the j-th column. With reference to A, these elementary operations are naturally equivalent to the addition of a suitable polynomial multiple of the i-th row of A to each of the other rows, and similarly for the columns.

Let us now remember that $m-1 > d$; hence, from the generality of the coefficients of the polynomial elements of A, we deduce that the hypersurfaces of S_d associated with the differential polynomials $\alpha_{12}, \alpha_{13}, \ldots, \alpha_{1m}$ have no point in common. Therefore, in virtue of the algebraic result mentioned above, there exist differential polynomials $\lambda_2, \lambda_3, \ldots, \lambda_m$ such that

$$\lambda_2\alpha_{12} + \lambda_3\alpha_{13} + \cdots + \lambda_m\alpha_{1m} = 1 .$$

Consequently, we can reduce the first element of the principal diagonal of the matrix A to unity, by adding to the first column the 2-nd, 3-rd, ..., m-th columns multiplied by $\lambda_2(1-\alpha_{11}), \lambda_3(1-\alpha_{11}), \ldots, \lambda_m(1-\alpha_{11})$

respectively. We then follow this with the reduction noted above, and so reduce the other elements of the first row and first column of A to zero.

Applying this sequence of operations k times, we obtain a matrix

$$B = \begin{Vmatrix} I_k & O \\ O & C \end{Vmatrix},$$

where I_k is the unit matrix of order k, the matrices O are zero matrices, and C is a $(n-k) \times (m-k)$ matrix whose elements are differential polynomials. Whence the first k equations of the system (6) are of the form

$$z_i = v_i \qquad (i = 1, 2, \ldots, k),$$

so that $z_1, z_2, \ldots, z_k$ are determined uniquely by differentiation only; whilst the remainder of the equations constitute a system of $n-k$ partial differential equations in $m-k$ unknowns.

By means of an analogous procedure, we can prove similarly that:

The general integral of a system of n linear partial differential equations with constant coefficients, in m unknown functions and d variables, can, in general, be obtained by operations of differentiation alone, when $m > d+1 \geqq n$ or $n > d+1 \geqq m$. In the first case, such an integral depends on $m-n$ arbitrary functions in the variables; whilst, in the second case, it is uniquely determined when the known functions, which appear as the right-hand sides of the equations of the given system, satisfy a suitable set of $n-m$ (differential) conditions of integrability.

§ 70. The solution of a system of homogeneous equations. — We now consider a system of homogeneous partial differential equations of the type

$$\sum_1^m {}_j \alpha_{ij} y_j = 0 \qquad (i = 1, 2, \ldots, m-1), \quad (35)$$

with $m-1$ equations in m unknowns. We denote by β_l the differential polynomial

$$\beta_l = (-1)^{m-l} \det \|\alpha_{ij}\|_{i=1,\ldots,m-1;\, j=1,\ldots,l-1,l+1,\ldots,m} \qquad (l = 1, 2, \ldots, m), \quad (36)$$

and observe that these differential polynomials satisfy the identities

$$\sum_1^m {}_j \alpha_{ij} \beta_j = 0 \qquad (i = 1, 2, \ldots, m-1). \quad (37)$$

We shall say that the β_l's are mutually prime, when their associated polynomials in $x_1, x_2, \ldots, x_d$ are mutually prime. In this case, the operators cannot all be identically zero, and we can establish the following theorem.

If the β_l*'s given by* (36) *are mutually prime, then the most general solution of the system* (35) *can be expressed in the form*

$$y_j = \beta_j z \qquad (j = 1, 2, \ldots, m), \quad (38)$$

where z *is an arbitrary function of* $x_1, x_2, \ldots, x_d$.

From (37) it is evident that the functions given by (38) are a solution of the system (35). We need therefore only prove the converse, i. e. show that any solution of (35) can be written in the form (38) for a suitable choice of the function z. The theorem is thus an immediate consequence of the following lemma:

If the operators β_l *defined by* (36) *are mutually prime, then the conditions of integrability of the system* (38) — *considered as differential equations in the single unknown* z — *are given by* (35) *and the relations which can be deduced from them by differentiation.*

From the considerations of § 67, p. 140, we deduce that the most general relation which can be obtained from (38) by differentiation and linear combination is of the form

$$\sum_1^m {}_j\, \lambda_j\, \beta_j\, z = \sum_1^m {}_j\, \lambda_j\, y_j\,,$$

where the λ_j's are differential polynomials. Whence the conditions of integrability of (38) are expressed by

$$\sum_1^m {}_j\, \lambda_j\, y_j = 0\,, \quad (39)$$

where the λ_j's are not all zero and identically satisfy

$$\sum_1^m {}_j\, \lambda_j\, \beta_j = 0\,.$$

But, since β_l is given by (36), the λ_j's are any set of differential polynomials such that the determinant of the following equations in μ is zero:

$$\lambda_j\, \mu_0 + \sum_1^{m-1} {}_i\, \alpha_{ij}\, \mu_i = 0 \qquad (j = 1, 2, \ldots, m)\,. \quad (40)$$

For any such choice of the λ_j's, equations (40) are satisfied by a suitable set of differential polynomials $\mu_0, \mu_1, \ldots, \mu_{m-1}$ which are not all zero, and which are mutually prime. In particular, $\mu_0 \neq 0$; for, otherwise, all the β_l's would be identically zero (from (36)), and this has been excluded in the enunciation of the lemma.

We now show that μ_0 *must be a (non-zero) constant.* Indeed, if this were not the case, there would be a hypersurface associated with μ_0. Then, by (40), either the generic point of any irreducible component of this hypersurface would also belong to the hypersurfaces associated with $\mu_1, \mu_2, \ldots, \mu_{m-1}$ (which in this case must all exist), so that the $\mu_0, \mu_1, \ldots, \mu_{m-1}$ would not be mutually prime, or such a generic point would belong to the hypersurfaces associated with $\beta_1, \beta_2, \ldots, \beta_m$, in which case the β_j's would not be mutually prime. However, with the present hypotheses neither situation can arise.

Consequently, we can define the differential polynomial ν_i by

$$\nu_i = -\frac{\mu_i}{\mu_0} \qquad (i = 1, 2, \ldots, m-1),$$

such that

$$\lambda_j = \sum_1^{m-1}{}_i\, \alpha_{ij}\, \nu_i \qquad (j = 1, 2, \ldots, m).$$

Whence we deduce that identically

$$\sum_1^{m}{}_j\, \lambda_j y_j = \sum_1^{m-1}{}_i\, \nu_i \sum_1^{m}{}_j\, \alpha_{ij}\, y_j,$$

so that the conditions of integrability, (39), of the system (38) are all consequences of (35). Therefore both the lemma and theorem are proved.

We observe that, amongst the conditions of integrability of (38), we have

$$\beta_i y_j = \beta_j y_i \qquad (i, j = 1, 2, \ldots, m).$$

These are automatically consequences of (35); but it is clear that, in general, the two systems are not equivalent. Thus, that the relations noted above are satisfied, does not suffice for the existence of a solution of (38). This is not contrary to theorem proved in § 67, p. 139, since the operators of the system (38) do not, in general, satisfy the condition (iii) of the enunciation of that theorem.

We can also prove the following generalization of the above theorem, by means of an argument which is perfectly analogous to the proof given above, and which can therefore be omitted.

Let

$$\sum_1^{m}{}_j\, \alpha_{ij}\, y_j = 0 \qquad (i = 1, 2, \ldots, n)$$

be a system of n homogeneous linear partial differential equations, with constant coefficients, in m unknowns $y_1, y_2, \ldots, y_m$ *and d variables* $x_1, x_2, \ldots, x_d$. *Suppose that* $n < m$, *and that the minors of order n extracted from the matrix* $\|\alpha_{ij}\|$ *are mutually prime, and so not all zero. If for example*

$$\beta = \det \|\alpha_{ij}\|_{i,j=1,2,\ldots,n} \neq 0,$$

then the most general solution of the system is given by

$$y_l = \sum_{n+1}^{m}{}_h\, \beta_{lh}\, z_h \qquad (l = 1, 2, \ldots, n),$$

$$y_h = \beta\, z_h \qquad (h = n+1, n+2, \ldots, m),$$

where β_{lh} *is the differential polynomial given by*

$$\beta_{lh} = (-1)^{n+1-l} \det \|\alpha_{ij}\|_{i=1,2,\ldots,n;\, j=1,\ldots,l-1,l+1,\ldots,n,h} \qquad (l = 1, 2, \ldots, n;\ h = n+1, \ldots, m),$$

and $z_{n+1}, z_{n+2}, \ldots, z_m$ *are* $m-n$ *arbitrary functions in the* x_i*'s.*

An immediate corollary of the first theorem of this section, can be enunciated as follows.

Let

$$\left.\begin{aligned} &\sum_1^m {}_j\, \alpha_{ij} y_j = 0 \qquad (i = 1, 2, \ldots, m-1) \\ &\sum_1^m {}_j\, \alpha_{mj} y_j = w \end{aligned}\right\} \tag{41}$$

be a system of m *linear partial differential equations in* m *unknowns* $y_1, y_2, \ldots, y_m$. *With the hypotheses of the first theorem of the present* § 70, *p.* 147, *the general integral of the system* (41) *is given by* (38), *where the* β_l*'s are calculated by* (36) *and, if* δ *denotes the determinant of the matrix* $A = \|\alpha_{ij}\|$, z *is a solution of the differential equation*

$$\delta z = w\,.$$

§ 71. The resolving system associated with a general set of m differential equations in m unknowns. — We now consider a system of n equations in $n-1$ unknowns which we write in the form

$$\sum_1^{n-1} {}_j\, \alpha_{ij} y_j = u_i \qquad (i = 1, 2, \ldots, n). \tag{42}$$

For brevity we put

$$\gamma_l = (-1)^{n-l} \det \|\alpha_{ij}\|_{i=1,\ldots,l-1,l+1,\ldots,n;\, j=1,2,\ldots,n-1} \qquad (l = 1, 2, \ldots, n). \tag{43}$$

With this notation we shall prove:

If the differential polynomials γ_l *given by* (43) *are mutually prime, then there exist functions* y_i *satisfying* (42) *if, and only if,*

$$\sum_1^n {}_i\, \gamma_i u_i = 0\,. \tag{44}$$

From the relations (43) we deduce that

$$\sum_1^n {}_i\, \alpha_{ij} \gamma_i = 0 \qquad (j = 1, 2, \ldots, n-1).$$

Hence (44) is immediately seen to be necessary.

In order to prove the sufficiency of this condition, we note that, from the considerations of § 67, p. 140, all the conditions of compatibility of the system (42) can be written in the form

$$\sum_1^n{}_i \lambda_i u_i = 0, \tag{45}$$

where the λ_i's are differential polynomials (not all zero), which are related only by

$$\sum_1^n{}_i \alpha_{ij} \lambda_i = 0 \qquad (j = 1, 2, \ldots, n-1). \tag{46}$$

By hypothesis, the γ_i's are not all identically zero; consequently, from (46), we deduce that the vectors λ and γ are proportional. Hence

$$\lambda_i = \varrho\, \gamma_i \qquad (i = 1, 2, \ldots, n),$$

where ϱ is a suitable differential polynomial, since the γ_i's are assumed to be mutually prime. Therefore evidently

$$\sum_1^n{}_i \lambda_i u_i = \varrho \sum_1^n{}_i \gamma_i u_i,$$

so that every condition of compatibility (45) is in fact a consequence of (44).

Finally, we shall prove the following theorem:

Let

$$\sum_1^m{}_j \alpha_{ij} z_j = u_i \qquad (i = 1, 2, \ldots, m) \tag{47}$$

be a system of m linear partial differential equations, with constant coefficients, in m unknowns; and let the minors of order l extracted from the matrix

$$\| \alpha_{ij} \|_{i=1,\ldots,l;\ j=1,\ldots,l+1}$$

be mutually prime for $l = 1, 2, \ldots, m-1$. Then it is possible to construct a sequence of m linear partial differential equations with constant coefficients, each member of which contains only one unknown, such that the complete solution of the given system (47) can be obtained from a particular integral of each of the first $m-1$ equations successively, and the general integral of the m-th equation of the constructed sequence.

Since the theorem is trivial for $m = 1$, we can assume that $m \geqq 2$ and proceed by induction on m. We thus apply the theorem, in the case of $m-1$ equations in $m-1$ unknowns, to the system

$$\sum_1^{m-1}{}_j \alpha_{ij} \bar{z}_j = u_i \qquad (i = 1, 2, \ldots, m-1). \tag{48}$$

Hence, from this system, we can construct a sequence of $m-1$ linear partial differential equations, each of which contains one unknown. From a particular integral of each of these equations, solved successively, we can construct a particular integral $(\bar{z}_1, \bar{z}_2, \ldots, \bar{z}_{m-1})$ of (48). We wish now to show that, using this particular integral, we can construct an m-th equation, the general solution of which leads to the general solution of (47).

For this we introduce new unknown functions $y_1, y_2, \ldots, y_m$, by putting

$$z_1 = \bar{z}_1 + y_1, \quad z_2 = \bar{z}_2 + y_2, \quad \ldots, \quad z_{m-1} = \bar{z}_{m-1} + y_{m-1}, \quad z_m = y_m.$$

Substituting in (47), and using (48), we obtain the system (41), where

$$w = u_m - \sum_{1}^{m-1}{}_j \, \alpha_{mj} \bar{z}_j .$$

Hence, applying the final corollary of § 70, p. 149, we reduce this system to an equation in a single unknown, the general solution of which leads back to the general solution of (47).

Historical Notes and Bibliography

The work of the present Part is taken from a recent paper of B. Segre [121]. This is in effect an extension of an earlier note [106], by the same author, in which the usefulness of the isomorphism between the set of differential operators and the corresponding ring of polynomials was already clearly demonstrated. The same isomorphism has also been exploited for similar ends, but in a less concrete form, by Gr. C. Moisil, An. Ac. Rep. Pop. Române, s. A, t. 2 (1949), 467—473; cf. also W. Gröbner, Monatsh. Math. u. Phis., 47 (1939), 247—284.

Some of the results given in §§ 62—65, pp. 132—137, relative to the case of a system of differential equations (2) in one variable, had previously been obtained by Jordan [57] n. 141, and by Picone and Fichera [83] n. 102 in the non-degenerate case, and by Picone and Ghizzetti [84] in the degenerate case. We recall that, with the notation of § 62, p. 133, the non-degenerate case is given by $n = m$ and det $A \neq 0$. The results concerning the homogeneous case had already been partially obtained by Gr. C. Moisil, Gaz. Mat. şi Fiz., 3 (1951), 1—24.

The literature concerning linear partial differential equations in several variables is very ample, and we shall return extensively to this subject in Part Nine. Here we shall confine ourselves to pointing out that the final result of § 68, p. 144, had already been given, for the simplest case of $d = 2$, by Sobrero [149]. But this treatment, which is substantially different from that of § 68, p. 142, seems to be defective, since it leads to the statement that the number of arbitrary constants appearing in the general integral of the system is equal to the product of the orders of the characteristic polynomials, i. e., with the present geometric interpretation, that it is equal to the total number of intersections (finite and infinite) of the associated plane curves; on this subject, cf. instead the final theorem of § 68, p. 144.

Also the corollary to the theorem of § 70, p. 149, when $w = 0$ had previously been found in the case $d = m = 2$ by Sobrero [150, § 2], and in the case $d = 2$, $m > 2$ by Fichera [47]. The present result demonstrates the existence of a potential function, in the sense of Sobrero [150, § 1], on quite general hypotheses. In the above work of Fichera [47], it is shown by an example that the result cannot always be extended to the case $d > 2$. This is, however, not opposed to the present result, since the example given has a certain particularity and does not satisfy the hypotheses stated here; indeed, this example merely affirms the necessity of our hypotheses. An interesting connection between systems of partial differential equations and generalized potential functions can also be found in GR. C. MOISIL, Bul. St. Ac. Rep. Pop. Române, 1 (1949), 341—351.

We conclude these notes, by recalling that there exists a marked analogy between the theory of partial differential equations, restricted to equations which are algebraic in the unknowns and their derivatives, and the theory of algebraic equations (cf. RITT [87, 88]). However, in the present rather restricted case of linear differential equations with constant coefficients, the analogy goes much further. Thus, as we have demonstrated, the possibility of solving a great many problems can be reduced to questions of a purely algebraic character.

Part Nine

Projective Differential Geometry of Systems of Linear Partial Differential Equations

§ 72. r-osculating spaces to a variety. — In S_n — over a field F which will generally be either the real or the complex one — consider a surface Φ defined as the locus of points x whose projective coordinates $(x^{(0)}, x^{(1)}, \ldots, x^{(n)})$ are functions of two parameters u and v. These functions will all be supposed continuous with derivatives of sufficiently high order. x_u will denote the point $\left(\frac{\partial x^{(0)}}{\partial u}, \frac{\partial x^{(1)}}{\partial u}, \ldots, \frac{\partial x^{(n)}}{\partial u}\right)$ and other derivatives of x will be similarly denoted; x_u and x_v are called the *first derived points of* x. The *tangent plane* at x to Φ is the plane joining the independent points x, x_u, x_v.

More generally, if V_k is a k-dimensional variety, the *tangent* S_k to V_k at a generic point x is the S_k determined by x and its k first derived points with respect to the k essential parameters to which V_k is referred. A *tangent* S_p to V_k at x is then, for $p < k$, an S_p through x and contained in the tangent S_k and, for $p > k$, an S_p containing the tangent S_k.

The *r-osculating space* at x to V_k is the smallest linear space containing x and all its derived points up to and including those of order r (cf. § 50, p. 108 and § 56, p. 122). The 2-osculating spaces are called *osculating spaces* and, for $k = 1$, the S_r which is r-osculating to a curve at a conveniently general point is also called, as usual, the osculating S_r at the point. The r-osculating space can also be defined as the smallest space containing each of the S_r which osculate, at x, the generic curves through x on V_k. But, for $r > 1$, such an r-osculating space is not in general covered by these osculating S_r.

From a certain value of r onwards, the r-osculating spaces will coincide with S_n. If r is less than this value, then each hyperplane through the r-osculating space meets V_k in a section for which x has at least multiplicity $r + 1$.

The osculating space to a surface Φ in S_n at one of its points x is determined by the six points

$$x, x_u, x_v, x_{uu}, x_{uv}, x_{vv}.$$

Therefore, if $n \geqq 5$, the osculating space has, in general, dimension 5. For a V_k in S_n with sufficiently large n, an r-osculating space to V_k has, in general, dimension $\binom{k+r}{r} - 1$. However, it can happen that the dimension is lower for each generic point by, say, i. This occurs when x and its successive derived points up to those of order r are connected by i linearly independent linear relations. In that case, the coordinates of a point which describes V_k correspond to some solutions of a system of i linearly independent linear equations in the partial derivatives of order r. V_k is then said to *represent* the system of equations; alternatively, V_k is called a V_k *integral* of the system.

A variety V_k is not, in general, projectively determined by such a system of partial differential equations. However, the introduction of a V_k representing such a system is useful for its study, since there are properties common to *all* integrals of a given system — as will be seen already in the case $k = r = 2$.

The special case of the dimension of the r-osculating space being lower than the maximum cannot occur for $r = 1$. For any r, each system of the above type represented by a V_k must consequently be such that no equation of the first order may be deduced from it. Hence

$$i \leqq \binom{k+r}{r} - (k+1)\,.$$

§ 73. Surfaces representing Laplace equations. — Consider the case $k = r = 2$; then $i \leqq 3$. If $i = 3$, the three points x_{uu}, x_{uv}, x_{vv} — and consequently also all the derived points — can be expressed as linear combinations of x, x_u, x_v; therefore, the plane of these three points contains each curve of the surface through x; and the surface in question is a plane. If $i = 2$, the surface lies in S_3 or is a developable. This brings us to the case $i = 1$, to which some attention will now be devoted.

Let Φ be such a surface and

$$A x_{uu} + 2\,B x_{uv} + C x_{vv} + 2\,D x_u + 2\,E x_v + F x = 0 \qquad (1)$$

the corresponding LAPLACE *equation*.

Over the real field, the equation is traditionally called *elliptic*, *parabolic* or *hyperbolic* according as $B^2 - AC$ is less than, equal to or greater than zero. Over the complex field, it is *parabolic* or *hyperbolic* as $B^2 - AC$ is or is not zero.

§ 74. The hyperbolic case. — To facilitate our discussion of Φ, we will define systems of curves, called nets, on Φ.

Let Ω be a non-singular quadric in S_3. If d is a tangent at P and d' the tangent at P polar to d, then d and d' are called *conjugate tangents*. Take on Ω a point P' near P. The tangent planes at P and P' to Ω meet in the line S' polar to PP'. As P' tends to P so that PP' tends to d, S' tends to d'.

Suppose that on Ω there are two families of curves so that, in a region R of Ω, through each point passes one curve of each family. Then the two families trace a *net* on Ω, if the two tangents to the curves at each point of R are conjugate tangents to Ω. Then we have the following theorem:

The tangent to the curves of one of the families of a net on Ω, at the points where these curves meet a curve γ of the other family, are the generators of a developable, i.e. they are tangents to a curve γ' (varying with γ), or they pass through a fixed point.

Now consider, on a surface Φ of S_n, two families of curves such that, in a region R on Φ, through each point P there is a curve of each family. If the curves of one family have the property that their tangents, at the points where these curves meet each curve of the other family, form a developable, then these two families constitute a *net* on Φ. This definition does not depend on the family chosen, meaning that, if the developables exist for one family, then they exist for the other.

Suppose that, on Φ, the curves with coordinates u and v form a net. Then the projective coordinates

$$x^{(i)} = f_i(u, v), \qquad i = 0, 1, \ldots, n,$$

have a special property. Namely, they satisfy an equation of the form

$$x_{uv} + a x_u + b x_v + c x = 0 \tag{2}$$

where a, b, c, are functions of u and v. Conversely, if the projective coordinates $x^{(i)}$ of the generic point x on Φ satisfy such an equation, then the curves with coordinates u and v trace a net on Φ.

Now take on Φ any two families of curves, which can be represented by the equations

$$f(u, v, \alpha) = 0, \quad g(u, v, \beta) = 0.$$

Suppose that, through each point of a region ϱ on Φ inside the region R where the net (u, v) is defined, there passes a curve of each family. Then α and β may be expressed as functions of u and v, and conversely. A new system of curvilinear coordinates is thus obtained and (2) takes the form

$$A x_{\alpha\alpha} + 2B x_{\alpha\beta} + C x_{\beta\beta} + 2D x_\alpha + 2E x_\beta + F x = 0. \tag{3}$$

If $A = C = 0$, the new curves also form a net. In any case, a *necessary condition for the surface Φ in S_n to have a net is that, when the surface is referred to any system of curvilinear coordinates (α, β), the projective coordinates of a generic point x satisfy a hyperbolic equation of the form* (3). The two families of curves $u(\alpha, \beta) = K$, $v(\alpha, \beta) = K'$ which comprise the net — if it exists — are the two families of integrals of

$$A\, d\beta^2 - 2B\, d\alpha\, d\beta + C\, d\alpha^2 = 0. \tag{4}$$

These two families, called *characteristics*, are distinct if $B^2 - AC \neq 0$ (over the real field, we naturally require that $B^2 - AC > 0$). Then, *a sufficient condition for the surface Φ in S_n to have a net is that the generic* *point x, referred to a system of coordinates (α, β) satisfies an equation* (3) *with distinct characteristics.*

The coordinates $x^{(0)}, x^{(1)}, x^{(2)}, x^{(3)}$ of a point x in S_3 which describes a surface Φ satisfy an infinity of equations (3). The most general can be written $L + \lambda L' = 0$ where $L = 0, L' = 0$ are two of them and λ is an arbitrary function of α and β. So *every surface in S_3 has an infinity of nets.* Providing $L = 0$ and $L' = 0$ do not have a family of characteristics in common, λ may be chosen so that one of the families of characteristics of $L + \lambda L' = 0$ is a family of given curves.

A surface Φ in S_4 generally satisfies just one equation (3). In this case, *if the characteristics are distinct, Φ has exactly one net.* There are just two possible exceptions: either the characteristics (4) of (3) coincide and Φ has no nets, or the $x^{(i)}$ satisfy two equations of the form (3) and Φ is a developable, with an infinity of nets.

On a surface Φ of S_n, with $n > 4$, there does not in general exist a net. The exceptions are twofold; either Φ satisfies just one equation (3) with distinct characteristics and has just one net, or it satisfies more than one equation (3) and has an infinity of nets. In the latter case, Φ is a developable. In fact, *every surface Φ in S_n, $n > 3$, which satisfies two equations of the type* (1) *is a developable or, as a special case, a cone.*

The net described by x is denoted by (x) or by (u, v).

Equation (2) is the canonical form for the hyperbolic LAPLACE equation; this form is chosen since its characteristics are given simply by

$$du\, dv = 0.$$

Let $h = a_u + ab - c$, $k = b_v + ab - c$. If $h = 0$ or $k = 0$, (2) can be reduced respectively to

$$x_{uv} + b x_v = 0 \quad \text{or} \quad x_{uv} + a x_u = 0.$$

In any case, under the transformation $x = \lambda x'$, where λ is an arbitrary, sufficiently differentiable function of u and v, h and k are absolute invariants of equation (2).

For an arbitrary transformation on the parameters α, β, the invariants h, k are simply multiplied by the Jacobian of the transformation. A necessary and sufficient condition for (2) to be reducible to an equation of the form

$$x_{uv} - m x = 0$$

by the above transformation is that $h = k$.

Let Φ be a surface in S_n, (x) a net on it and (2) the corresponding Laplace equation. The tangents to the curves v of the net, at the points where these curves meet a given curve u, form a developable or a cone. These lines $x x_v$ remain, for $v = K$, tangents to a certain curve which generally varies with the constant K, or they pass through the same point, also variable with K.

The coordinates of the point x_1 of xx_v, where this line touches the edge of regression, or which remains fixed, are given by

$$x_1 = x_v + a x.$$

As u and v vary, x_1 describes a net (x_1) and we will say that (x_1) is the LAPLACE *transform* of (x) (and, correspondingly, that the surface Φ_1 described by x_1 is the LAPLACE transform of Φ) in the direction of the curves v, providing $h \neq 0$. Similarly, providing $k \neq 0$, the point x_{-1} defined by

$$x_{-1} = x_u + b x$$

describes the net (x_{-1}), the LAPLACE transform of (x) in the direction of the curves u; (x_{-1}) lies on the surface Φ_{-1}.

Consider the nets (x) and (x_1); the line xx_1 generates a congruence with special properties. For $u = u_0$, x describes a curve v of the net (x) to which xx_1 is tangent. So xx_1 traces a developable (u_0) which belongs to the congruence. The tangent plane to this developable along the generator xx_1 contains the tangent at x_1 to the curve $u = u_0$ of the net (x_1) and is the tangent plane at x_1 to Φ_1; it is also the osculating plane at x of the curve $u = u_0$ of the net (x). Further, the developable (u_0) is circumscribed to Φ_1 along the curve $u = u_0$.

As u_0 varies, one has an infinity of developable surfaces, whose generators constitute the congruence (xx_1). All these developables are circumscribed to Φ_1 along the curves v of the net (x_1) and their edges of regression are the curves v of the net (x). Similarly, there is another family of developables circumscribed to Φ along the curves u of (x), whose edges of regression are the curves u of (x_1) and whose generators also constitute the congruence (xx_1).

The congruence (xx_{-1}) has similar properties and the two congruences (xx_1) and (xx_{-1}) are called *congruences of the net.*

If (x_1) is not reduced to a curve, a new net (x_2) may be formed in the same way; it is called the *second* LAPLACE *transform of the net* (x) *in the direction of the curves* v. Hence we obtain a series, generally infinite,

$$(x_1), (x_2), \ldots, (x_m), \ldots.$$

Similarly, in the direction of the curves u, we obtain the series

$$(x_{-1}), (x_{-2}), \ldots, (x_{-m}), \ldots.$$

Correspondingly, we have two series of surfaces which, written together, become

$$\ldots \Phi_{-m}, \ldots, \Phi_{-2}, \Phi_{-1}, \Phi, \Phi_1, \Phi_2, \ldots, \Phi_m, \ldots. \quad (5)$$

There are four possibilities for the series (5). It may be infinite on both sides, terminate on either side and not the other, or terminate on both sides. All possibilities actually occur.

§ 75. The parabolic case. — The canonical form for a *parabolic* LAPLACE *equation* is

$$x_{uu} + a x_u + b x_v + c x = 0. \tag{6}$$

Let S_n, $n \geqq 3$, be defined over the real field and take, as in § 72, a system of projective coordinates $(x^{(0)}, x^{(1)}, \ldots, x^{(n)})$. Then the equations

$$x^{(i)} = x^{(i)}(u, v), \qquad i = 0, 1, \ldots, n,$$

where the right hand sides are $n+1$ linearly independent regular solutions of (6), represent a differentiable variety of S_n — in general a surface Φ called an integral surface of (6). On Φ the characteristics $v = K$ constitute a system of *asymptotics* in that, at each of their points, the corresponding tangents have at least double contact with Φ, which is tantamount to saying that the osculating plane coincides with the tangent plane of the surface. For these asymptotics to be lines, it is necessary and sufficient that $b = 0$ in (6). In such a case, Φ is a ruled surface and (6) reduces essentially to an ordinary differential equation.

If $n > 3$, the asymptotics $v = K$ on Φ are not lines and are uniquely defined by Φ (for $n = 3$, Φ may contain two different systems of asymptotics). Such an ∞^1 system of curves will be denoted by τ and called a *weave*. Its curves have ∞^2 tangents, which constitute a *parabolic congruence* (with *focal surface* Φ and *focal weave* τ) and which fill a variety $W_3 = W$, called a *parabolic ruled variety*.

More generally, if s denotes a positive integer, not too large compared with n, one can consider the variety W_{s+2} generated by the ∞^2 osculating spaces S_s to the curves of τ.

One can show that, if τ is any weave drawn on a surface Φ of S_n, each curve of τ in general belongs to the S_n. More precisely, we have the following theorem:

Given any integer h satisfying $h \geqq 1$ and $2h \leqq n+1$, then, at each generic point x of Φ, the h-osculating space has in general exact dimension $2h$ and coincides with the S_{2h} that osculates at x the curve of τ through x. The only exception is the particular case in which, for $h > 1$, the curves of τ lie in spaces S_{2h-1} forming a developable (including the possibility that all these S_{2h-1} coincide with S_n), and that at each generic point of Φ, the h-osculating space coincides with the corresponding S_{2h-1}.

Further, if τ is general and s satisfies $1 \leqq s \leqq n-2$, the variety W_{s+2} generated by the ∞^2 S_s osculating τ admits a fixed tangent S_{s+2} along each generator. The ∞^2 tangent spaces thus defined are in fact the osculating S_{s+2} of the curves of τ.

Given a weave τ, it is of interest to determine in some intrinsic way a system of curves, traced on the surface Φ containing τ and locally unisecant to the curves of τ. Such a system is called a *warp* with respect to τ. By a transformation $\bar{u} = g(u, v)$, $\bar{v} = v$ with $g_u \neq 0$, it can be reduced to $u = K$. This induces a corresponding *normalization* of the LAPLACE equation.

If the curves of τ lie in spaces S_{2h+1} of odd dimension, one obtains a warp with respect to τ consisting of curves of Φ at each point of which there is an osculating S_{h+1} lying in the S_{2h} which is the h-osculating space to Φ.

If the curves of τ lie in spaces S_{2h} of even dimension greater than two, one obtains a warp with respect to τ consisting of curves of Φ at each point of which there exists an osculating S_{h+1} meeting the $(h-1)$-osculating S_{2h-2} to Φ in an S_h.

§ 76. Surfaces representing differential equations of arbitrary order. — With the integers k, r, i introduced in § 72, we have studied the case $k = r = 2$, $i = 1$. We are naturally led to consider the case of surfaces with equations of any order ($k = 2$, r arbitrary) and the case of varieties of any dimension with equations of the second order (k arbitrary, $r = 2$).

In the former case, it may be observed that, if the dimension of the r-osculating and $(r+1)$-osculating spaces at each generic point of a surface differ by one — and so are ϱ and $\varrho+1$ respectively — then the surface contains ∞^1 curves in the $S_{\varrho-r}$ of an ordinary developable (possibly degenerate) or else it lies in $S_{\varrho+1}$. (An ∞^1 of spaces S_h is an *ordinary developable of kind h* when two infinitely close S_h lie in an S_{h+1}. It consists, in general, of the S_h osculating a curve — the case which has occured several times above — or, in the degenerate case, of the S_h which are the projections from a fixed S_g of the S_{h-g-1} osculating a curve in S_n where $0 \leq g \leq h-1$; the latter case will not, henceforth, be excluded).

Hence it follows that the surfaces representing a system of order r with

$$i = r(r-1)/2 + h$$

($h > 0$) linearly independent equations either consist of ∞^1 curves in the S_{r-h} of an ordinary developable or lie in S_{2r-h}. For values of i with $0 \geqq h \geqq -5$, the corresponding class of surfaces has also been determined.

§ 77. Varieties of arbitrary dimension representing Laplace equations. — Now consider the case of k arbitrary and $r = 2$. Then, if

$$i = k(k-1)/2 + l$$

with $l > 0$, it may be shown that the V_k lies on a variety U_q, the locus of ∞^h spaces S_p such that the tangent S_q at the points of an S_p lie in an S_{2k-h-1} with $0 \leq h \leq k-l$.

If x is the point tracing V_k, a generic point y of the variety W generated by the tangent S_k to V_k is given by

$$y = x + \lambda_1 x_1 + \cdots + \lambda_k x_k$$

where the lower suffixes of the x_i denote derivatives of x with respect to the essential parameters $t_1, \ldots, t_k$ of V_k. Since the system consisting of y and its first derivatives with respect to the $2k$ parameters reduces to

$$x, x_1, \ldots, x_k, \lambda_1 x_{11} + \ldots + \lambda_k x_{1k}, \ldots, \lambda_1 x_{k1} + \cdots + \lambda_k x_{kk},$$

it follows that, if V_k represents $i > k(k-1)/2$ linearly independent LAPLACE equations, then W has dimension less than that one, namely $2k$, which occurs for V_k embedded in a space of sufficiently large dimension. Precisely, if $i = k(k-1)/2 + l$, then the dimension of W is at most $2k - l$. Less obvious, however, is the converse: namely, if W has dimension $2k - l$, then the V_k represents exactly $i = k(k-1)/2 + l$ lineary independent LAPLACE equations — with a single exception. This exception occurs when the system of associated quadratic forms has an apolar system with its Jacobian matrix of rank $k - l$.

§ 78. Generalized developables. — Take an ordinary V_{k+1} locus of ∞^1 S_k. Then the tangent S_{k+1} at the points of a generator S_k correspond projectively to their points of contact. However, this projectivity may be degenerate; and this happens when there exists a *singular* or *focal* space $[k_1]$ common to the generator S_k considered and its first successive S_k. If this happens for each generic S_k of the family, then the family is called a *developable*. Excluding the case that the ∞^1 spaces S_k have a linear space in common, suppose that, on the variety $V^{(1)}$ generated by the ∞^1 spaces $[k_1]$, each generator has a singular space $[k_2]$. Hence we obtain a variety $V^{(2)}$ of ∞^1 spaces $[k_2]$; and so on until we reach the variety $V^{(\nu)}$ whose generic generators $[k_\nu]$ have no singular point. Each of the spaces $[k]$, $[k_1], \ldots, [k_\nu]$ may be thought of either as the meet of two successive ones of the preceeding system or as the join of two successive ones of the succeeding system — where, of course, the required system exists. Thus, on each generator $[k]$ of V, there is a $[k_1]$, a $[k_2], \ldots,$ a $[k_\nu]$, the respective intersections of $2, 3, \ldots, \nu + 1$ successive generators of V; and each $[k]$ of V contains $\nu + 1$ successive $[k_\nu]$ of $V^{(\nu)}$.

For the above situation to exist, the following inequalities form a necessary condition:

$$n - k \geqq k - k_1 \geqq k_1 - k_2 \geqq \cdots \geqq k_{\nu-1} - k_\nu \geqq k_\nu + 1.$$

For a non-degenerate ordinary developable, the numbers $k_1, k_2, \ldots$ have the respective values $k-1, k-2, \ldots, 1, 0$.

These considerations can be generalised in S_n to a $V_{k+\alpha}$, the locus of ∞^α spaces S_k with $\alpha + 2k \leqq n$. Then, the ∞^k spaces $[k+\alpha]$, which are tangent to $V_{k+\alpha}$ at the points of a fixed generator G, and the $\infty^{\alpha-1}$ spaces $[2k+1]$ each of which joins G to an infinitely close S_k of $V_{k+\alpha}$ simply cover the same $V_{2k+\alpha}$, which is an algebraic cone with vertex G. The order of $V_{2k+\alpha}$ is $\binom{k+\alpha-1}{k}$ and its class is $\binom{k+1}{\alpha-1}$ or $\binom{\alpha}{k}$ according as $\alpha \leqq k+1$ or $\alpha \geqq k+1$. $V_{2k+\alpha}$ may also be defined as the cone which projects, from G, the infinitely near points of G on $V_{k+\alpha}$. $V_{2k+\alpha}$

is in general embedded in a space of dimension $\alpha k + \alpha + k$ providing that this number is not greater than n.

Singular points exist on the generators of $V_{k+\alpha}$ generally only when $2k + \alpha > n$. They form a variety of dimension $2k + \alpha - n - 1$ and order $\binom{n-k}{\alpha-1}$ unless $k + \alpha > n$, in which case they include all the points of a generator.

The notion of ordinary developable can be extended in several other ways.

§ 79. Varieties of arbitrary order representing differential equations of arbitrary order. — To continue the development of r-osculating spaces to a V_k from § 72, consider the $\binom{k+r}{r}$ derived points of x on V_k (with essential parameters $t_1, \ldots, t_k$) up to those of order r. There may be a linear relation

$$\sum a_{i_1 i_2 \ldots i_m} x_{i_1 i_2 \ldots i_m} = 0 \tag{7}$$

between them, where $x_{i_1 i_2 \ldots i_m}$ denotes — as in § 77 — the derived point with respect to the parameters $t_{i_1}, t_{i_2}, \ldots, t_{i_m}$.

When the equation has order r and not less, the sum of the terms of order r in (7), viz.

$$\sum a_{i_1 \ldots i_r} x_{i_1 \ldots i_r} \tag{8}$$

is called the *principal part* of (7). Denote by ϱ_r the maximum number of linearly independent linear equations (7) satisfied by V_k modulo the equations of order less than r. Then, if d_r is the dimension of the r-osculating space S_{d_r} at a generic point x of V_k,

$$d_r = \binom{k+r}{r} - 1 - \sum_{i=1}^{r} \varrho_i. \tag{9}$$

Putting

$$e_r = d_r - d_{r-1} - 1,$$

gives

$$e_r = \binom{k+r-1}{r} - \varrho_r - 1.$$

Hence, if C is any curve on V_k which passes r-regularly through x (i.e., any $r+1$ distinct points of the neighbourhood of x on C are linearly independent), the osculating S_r of C at x lies in S_{d_r}. For $r = 1$, $S_{d_1} = S_k$ is covered by the ∞^{k-1} tangent lines $S_r (= S_1)$ obtained as C varies arbitrarily through x. For $r \geqq 2$, these S_r generate, in general, a proper subvariety of S_{d_r}.

Let P, Q denote respectively the $(r-1)$- and the r-osculating spaces of V_k at x. P and Q have dimensions d_{r-1} and d_r; P lies in Q and $d_{r-1} \leqq d_r$. The *special case* $P = Q$ for a general x can only occur when $r \geqq 2$. Then P and Q do not depend on the x chosen and such a P (coinciding with Q) is the smallest projective space containing V_k. As this is S_n, $n = d_{r-1} = d_r = d_{r+1} = \ldots$.

If this special case is ruled out so that $e_r \geqq 0$, we may consider the projective space Σ_{e_r} determined by the star defined by P and Q. The "points" and "hyperplanes" of Σ_{e_r} are respectively the $(d_{r-1}+1)$-dimensional and the $(d_r - 1)$-dimensional spaces of Q containing P.

If $r = 1$, then $e_1 = k - 1$ and Σ_{e_1} is the projective space whose "points" are the ∞^{k-1} tangent lines p of V_k at x and, consequently, have $dt_1, \ldots, dt_k$ as projective coordinates. If $r \geqq 2$, the osculating S_r at x to the curves C on V_k through x are joined to P by spaces of dimension $d_{r-1} + 1$ whose locus is therefore a cone K with vertex P; this cone will be called the Del Pezzo *cone of index k* of V_k at x.

A general generator of K depends only on the tangent line p cut out by the corresponding S_r on the tangent S_k to V_k at x (i.e., it remains fixed as C varies on V_k providing C touches p at x). Hence *every* Del Pezzo *cone is unirational.* The rational transformation between $\Sigma_{e_1} = \Sigma_{k-1}$ and K (considered as a variety of Σ_{e_r}) transforms the ∞^{e_r} "hyperplane" sections of K into a linear system Φ_r of dimension e_r, which consists of "forms" of Σ_{k-1} — algebraic cones of S_k with vertex x. The forms of Φ_r are of order r and can also be defined in the following manner.

Any hyperplane S_{n-1} of S_n passing through P but not containing Q cuts out on V_k a $(k-1)$-dimensional manifold W, which has at x an algebroid singularity of exact multiplicity r. The tangent cone of W at x is therefore a $(k-1)$-dimensional algebraic cone φ_r of order r lying in S_k and with vertex x. The cone φ_r depends only on the space cut out by S_{n-1} on Q. This S_{n-1} is in fact a "hyperplane" of Σ_{e_r} and, as it varies in Σ_{e_r}, φ_r describes the linear system Φ_r.

Let us now consider varieties V_k $(k \geqq 2)$ of S_n whose $(r-1)$- and r-osculating spaces have dimensions differing by one (cf. § 76); i.e., $e_r = 0$ for a certain $r \geqq 2$. This is equivalent to saying that V_k satisfies a system of

$$\varrho_r = \binom{r+k-1}{r} - 1$$

homogeneous linear partial differential equations of order r, with linearly independent principal parts.

The condition $e_r = 0$ means that the linear system Φ_r consists of a single form, φ_r say. Thus V_k is characterized as follows: a variable S_{n-1} of S_n through P but not containing Q cuts out on V_k a $(k-1)$-dimensional manifold, which admits at x a fixed tangent cone (coinciding with φ_r).

For completeness, two cases must be distinguished, according as — for a general point x of V_k — the form φ_r is or is not the r-th power of a linear form.

In the second case, it can be shown that $d_r = d_{r+1}$. Hence, from the "special case" earlier in this section, the space Q remains fixed as x varies on V_k. So Q coincides with S_n, V_k belongs to a space of dimension $n = d_r$ and, from (9), n has an upper bound.

In the first case, it can be shown that the Pfaffian form of which φ_r is now the r-th power is integrable on V_k and so defines a pencil of varieties W_{k-1}. Then the $(r-1)$-osculating space P at a general point x of V_k

must contain the whole variety W_{k-1} passing through x and must also touch V_k along this W_{k-1}. Also P remains fixed as x moves on W_{k-1}. Consequently, there are only ∞^1 positions for P and these spaces form a developable of kind d_{r-1}. The single varieties W_{k-1} are contained in the generators of a developable of kind $d_{r-1} - 1$ defined by the previous one. So, now there is no upper bound for n and *V_k is the locus of ∞^1 varieties U_{k-1} lying in the generators of a developable of kind $d_{r-1} - 1$.*

§ 80. The postulation of varieties by conditions on their r-osculating spaces. — An interesting, general problem is the following one:

Given $c (\geqq 2)$ positive integers $k_1, k_2, \ldots, k_c$, determine the varieties V_k with $k \geqq 2$ for which c generally chosen osculating spaces of indices $k_1, k_2, \ldots, k_c$ are linearly dependent.

It may be shown that the required V_k are those for which the locus generated by the join of c osculating spaces of indices $k_1 - 1, k_2 - 1, \ldots, k_c - 1$, as these osculating spaces vary arbitrarily, has dimension less than usual.

This problem is important in algebraic geometry in the case of an algebraic V_k, especially when V_k is a generalized VERONESE variety, i.e. (§ 50, p. 110), the projective image of the linear system consisting of all the forms of a given order m in S_k. In this case the problem reduces to that of deciding what values $k, m, k_1, \ldots, k_c$ must take in order that the conditions on the forms of order m in S_k through c assigned points with multiplicities $k_1 + 1, \ldots, k_c + 1$ are linearly dependent for any choice of the c points.

Let us take the general problem in the simplest case: $k = c = 2$, $k_1 = k_2 = 1$. We are required to determine the differentiable surfaces V_2 of S_n, every pair of whose tangent planes are incident. It suffices to consider $n \geqq 5$, since, for $n \leqq 4$, the condition is automatically satisfied. A classical result of DEL PEZZO shows that an *algebraic* surface satisfying the condition is *a cone or the* VERONESE *surface*. We shall now show that *the same holds true on supposing only that the surface has continuous derivatives of the first and second order.*

Let V_2 be such a surface in S_n with $n \geqq 5$ and suppose then that every pair of tangent planes to V_2 intersect. Take a general point x on V_2 and consider the curve C cut out on V_2 by a general hyperplane S_{n-1} of S_n through the tangent plane P to V_2 at x. The curve C cannot have at x a singular point of multiplicity greater than two. For, otherwise, P would also be the 2-osculating space of V_2 at x; hence, from the "special case" of § 79, V_2 would be contained in P and so could not belong to S_n. The curve C therefore has a double point at x, and its two tangents at x — as S_{n-1} varies through P — describe on P the linear system Φ_2.

Let us firstly consider the case when the two tangents remain fixed, i.e., Φ_2 has dimension $e_2 = 0$. Then, from § 79, V_2 lies in S_3 or is a ruled developable of the first kind (this is also a result of C. Segre). The former is impossible as $n \geqq 5$; the latter means that V_2 is a cone.

There remains the case that Φ_2 has dimension $e_2 \geqq 1$. Then, if one chooses one of the ∞^1 tangents to V_2 at x, say p, there exists a curve C — the intersection of V_2 with an S_{n-1} through P — which admits two distinct tangents at x, one of which is p. Denote by Γ the branch of C touching p at x, by x' a general point of Γ and by P' the plane touching V_2 at this point. Γ does not lie in P, as p is general.

By hypothesis, P and P' meet. They cannot meet in a line, since otherwise S_{n-1} would contain this line and the point x' not lying on it, and so would contain P'. Hence S_{n-1} would touch V_2 along Γ and C would then consist of Γ counted twice, whereas it has two distinct tangents at x. It follows that S_{n-1} meets P' in a well-defined line, which touches Γ at x' and contains the point y of intersection of P and P'. Hence Γ, having all its tangents incident to the fixed plane P, lies in an S_3 containing P. This S_3 therefore contains the osculating plane of Γ at x and so is a generator of the Del Pezzo cone K, of index 2, to V_2 at x.

As p describes the pencil of lines touching V_2 at x, the curve Γ describes a two-dimensional region of V_2; hence V_2 lies on K and, similary, lies on the Del Pezzo cone K^* at any point x^* off V_2.

It follows that K^* cannot be an S_4, since $n \geqq 5$. So the 2-osculating space of V_2 at x^* has dimension 5 — implying that $e_2 = 2$ — and K^* is an irreducible quadric cone of this space with vertex the tangent plane P^* to V_2 at x^*.

Now, K^* cannot contain S_3 as P and P^* do not meet in a line. Hence the ∞^2 quadric cones K^* meet S_3 in ∞^2 ordinary quadric surfaces and each of these quadrics must contain Γ. Hence Γ is necessarily on a line or a conic.

On the other hand, V_2 contains at least ∞^2 curves Γ, since there is an infinity of these curves through each point of V_2. Hence V_2 contains ∞^2 (pieces of) lines or conics. In the former case, V_2 would be a region of a plane, which is impossible. In the latter case, using a classical result of Darboux on surfaces containing a double infinity of conics, we deduce that V_2 is the Veronese surface. This concludes the proof of the theorem.

Historical Notes and Bibliography

A great deal of the foregoing Part — particularly §§ 72—74, 76—78 is based on Terracini's Appendix III to [48], where further references are to be found. Properties of surfaces representing hyperbolic Laplace equations are considered in depth by C. Segre [127] and by Tzitzeica [151]. The latter especially gives many details on the series of nets on surfaces obtained by successive Laplace transforms. This topic is considered as well by Bol [16], whose book is noteworthy for its vast bibliography; however, the methods used in this work owe very little to algebraic geometry. The treatment of the parabolic Laplace equation in § 75 is taken from

B. SEGRE [124]. Various types of generalized developables including those of § 78 where introduced by C. SEGRE [128]. §§ 79—80 on r-osculating spaces to a variety V_k are taken from B. SEGRE [123]. DEL PEZZO's result on algebraic surfaces, every two of whose general tangent planes intersect, is to be found in BERTINI [13] p. 393, where on p. 394 is DARBOUX's result that a surface containing a double infinity of conics is the VERONESE surface in S_5 or one of its projections.

Part Ten

Correspondences between Topological Varieties

Reverting to the considerations of the first two Parts, we shall now illustrate some aspects of the global theory of *correspondences* between varieties. Previously we have obtained results of a local character and these fit into the global theory in a natural manner; also, the results now obtained should be compared with those of Parts Five and Six. However, the present global theory replies to questions and uses methods differing greatly from those which have concerned us so far; so that, for example, some of the results which we shall obtain are valid for any topological variety, and not only for differentiable or analytic varieties.

§ 81. Products of topological varieties. — We first of all consider any *two topological varieties*, V and V', having the same dimension n, each of which is closed, connected and orientable. We denote by R_p and R'_p the BETTI numbers of V and V' respectively (for $p = 0, 1, \ldots, n$; and where $R_0 = R'_0 = 1$, $R_p = R_{n-p}$, $R'_p = R'_{n-p}$). For the general notions relative to the homology groups, the BETTI groups, the duality of POINCARÉ, etc., we refer the reader to B. SEGRE [122], chap. II, for example. We choose, in an arbitrary manner, the bases for the p-cycles of V and V' (relative to the homology with division), and denote them by

$$C^i_p, C'^j_p \quad (i = 1, 2, \ldots, R_p; j = 1, 2, \ldots, R'_p; p = 0, 1, \ldots, n).$$

Further, we indicate their KRONECKER indices by $a^{ii'}_p$, $a'^{jj'}_p$, where

$$a^{ii'}_p = [C^i_p, C^{i'}_{n-p}], \qquad a'^{jj'}_p = [C'^j_p, C'^{j'}_{n-p}],$$

and the matrices of these indices by A_p, A'_p. Thus A_p and A'_p are square and non-degenerate, with orders R_p and R'_p respectively. Naturally, however, if $R_p = 0$ or $R'_p = 0$, then we no longer consider such matrices. We note that they satisfy the relations

$$A_p = (-1)^{p(n-p)} (A_{n-p})_{-1}, \quad A'_p = (-1)^{p(n-p)} (A'_{n-p})_{-1}, \tag{1}$$

where the suffix -1 denotes transposition. Also, we can always choose

$$C^1_0 = P, \; C'^1_0 = P', \; C^1_n = V, \; C'^1_n = V',$$

where V and V' are the given varieties arbitrarily oriented, and P, P' denote points of V, V' respectively; whence we deduce that

$$A_0 = A_n = A'_0 = A'_n = \|1\|$$

(see, for example, B. SEGRE [122], chap. II).

It is well known that, on the *topological product* $W = V \times V'$, a BETTI base is given by the totality of products of the form $C^i_p \times C'^j_q$; whence we see that every n-cycle Γ_n on W satisfies a homology with division of the form

$$\Gamma_n \approx \sum_0^n{}_p\, x^p_{ij} (C^i_p \times C'^j_{n-p}), \tag{2}$$

where the summation with respect to i, j is understood. Thus the cycle defines a set of $n+1$ matrices

$$X^p = \| x^p_{ij} \|_{i=1,2,\ldots,R_p;\, j=1,2,\ldots,R'_p} \qquad (p = 0, 1, \ldots, n);$$

again, such a matrix is lacking if $R_p R'_p = 0$. We note, in particular, that each of X^0, X^n always reduces to a single element x^0_{11}, x^n_{11}, which we shall denote simply by α, α' respectively.

§ 82. Correspondences and relations. — In order to give a meaning to the concept of a correspondence between V and V', one could, from the viewpoint of set-theory, regard it as an arbitrary set of points on the product variety $W = V \times V'$. But, for simplicity, we prefer to use the word *relation* in this case, reserving the word *correspondence* — as is usual in topology — for a relation T between V and V' which is represented on W by an n-cycle $\Gamma = \Gamma_n$. The correspondence T can then be endowed with the same topological structure as is possessed by Γ: this we shall always do in the sequel. One effect of this procedure, is to distinguish T and $-T$ (represented by $-\Gamma$ on W), though T and $-T$ coincide when they are considered simply as relations between V and V'. An arbitrary correspondence T between V and V', with the above significance of the term, induces between the p-th BETTI groups H_p of V and H'_p of V' ($p = 0, 1, \ldots, n$) a *linear mapping*, which is intrinsically defined in the following manner. To each p-cycle, C_p, of V corresponds the p-cycle TC_p on V' given, to within a homology with division, by

$$TC_p = \mathrm{pr}_{V'}(\Gamma \cdot C_p \times V'), \tag{3}$$

where the right-hand side denotes the projection onto V' of the intersection (on W) of the cycles Γ and $C_p \times V'$.

The geometric significance of (3) is immediate. With the notation of the preceding section, we deduce the homologies

$$TC^i_p \approx \theta^i_{p,j} C'^j_p, \tag{4}$$

where the coefficients $\theta^i_{p,j}$, for a fixed integer p ($p = 0, 1, \ldots, n$), are the elements of a matrix

$$\Theta_p = \|\theta^i_{p,j}\|_{i=1,2,\ldots,R_p;\, j=1,2,\ldots,R'_p}$$

which can be expressed in terms of the matrices introduced in § 81, pp. 164—165, according to the relations

$$\Theta_p = (-1)^{p(n-p)} A_p X^{n-p} \qquad (5)$$

(cf. LEFSCHETZ [67] p. 268; or, for example, B. SEGRE [122], n. 86, pp. 229—232).

Thus, for the special case in which V and V' are superimposed, we can take $C'^i_p = C^i_p$, so that $A'_p = A_p$. We then denote by Δ the cycle of W which represents the *identical transformation* of V into itself; and we suppose Δ to be oriented so that, when it is projected onto V, the orientations of the two varieties correspond. Δ is thus given by the right-hand side of (2), when we put

$$X^p = (A_p)^{-1}_{-1} \qquad (p = 0, 1, \ldots, n). \qquad (6)$$

From (1) and (5) we see that this is equivalent to supposing that each Θ_p now reduces to a unit matrix (with order $R_p = R'_p$). Moreover, from intuitive considerations of (4), we see that this in fact should happen..

If $V = V'$, we define the *algebraic number*, $u = u(T)$, *of united points* of a correspondence, T, by means of the relation

$$u = [\Gamma, \Delta], \qquad (7)$$

where $[\Gamma, \Delta]$ is the KRONECKER index of Γ and Δ on W. We could equally put $u = [\Delta, \Gamma]$, which would have the effect of multiplying u by $(-1)^n$. The number of united points can be calculated by means of the equality

$$u = \sum_0^n {}_p (-1)^p \operatorname{tr} \Theta_p, \qquad (8)$$

where tr Θ_p denotes the trace of the square matrix Θ_p, when we have identified the cycles C'^i_p with the cycles C^i_p. This is the fundamental *formula of* LEFSCHETZ, and it can be easily deduced from the preceding results (cf. for example B. SEGRE [122], n. 89, pp. 233—245). For the case in which T is the identity, we see from (8) that the number u, as is well known, reduces to the EULER-POINCARÉ characteristic of V.

§ 83. **Inverse correspondences.** — The cycle $\Gamma = \Gamma_n$, which defines the correspondence T between V and V' as we have indicated in § 82, p. 165, defines, in a similar manner, a correspondence between V' and V; this we shall call the *inverse correspondence* of T, and we shall denote it by T^{-1}. More precisely, by interchanging V and V' (and similarly the cycles C and C'), we obtain, for T^{-1}, the following relation instead of (3):

$$T^{-1} C'_p = \mathrm{pr}_V (\Gamma \cdot V \times C'_p).$$

This formula already appears in HOPF [55] (formula (8b)).

On the right-hand side of this relation, it is now found convenient to interchange the two terms in brackets, in order that the equation analogous to (6) should be still valid when T is the identical transformation of V into itself, since then T manifestly coincides with T^{-1}. This change of the order means, at most, that we should change the sign of $T^{-1}C'_p$. We thus obtain the analogues of equations (3):

$$T^{-1}C'_p = \mathrm{pr}_V(V \times C'_p \cdot \Gamma). \tag{3'}$$

If also we put

$$T^{-1}C'_{pj} \approx \theta'^{j}_{p,i}C^i_p, \qquad \Theta'_p = \|\theta'^{j}_{p.i}\|_{j=1,2,\ldots,R'_p;\, i=1,2,\ldots,R_p}, \tag{4'}$$

then we deduce that

$$\Theta'_p = A'_p(X^p)_{-1} \qquad (p = 0, 1, \ldots, n), \tag{5'}$$

which is analogous to (5). From (5) and (5′), using also (1), we obtain

$$\Theta'_p = A'_p(\Theta_{n-p})_{-1}(A_p)^{-1}, \tag{9}$$

which shows that, with the present choice of signs, every Θ'_p must in fact be a unit matrix, of order $R_p = R'_p$, when V' is superimposed on V and the cycles C'_p are identified with the cycles C_p.

If $V = V'$, then the *algebraic number u' of united points* of the inverse T^{-1} of any correspondence T can be taken to be equal to the number u, given by (7), relative to T. Thus, from (9), we obtain the formula

$$u' = \sum_0^n {}_p(-1)^{n-p}\,\mathrm{tr}\,\Theta'_p, \tag{8'}$$

which, apart possibly from the sign, is the same as the formula of LEFSCHETZ (8).

§ 84. Homologous correspondences. — If, as in the previous sections, V and V' denote two given topological varieties, the totality of correspondences T (in the sense of § 82, p. 165) between V and V' constitute a *modulus*, which is isomorphic to that of the n-cycles on $W = V \times V'$. Consequently, we can speak — with an obvious meaning — of correspondences which are homologous (with or without division).

In view of the considerations in § 82, p. 165, every correspondence T between V and V' intrinsically defines a *linear mapping* of the p-th BETTI group, H_p, of V into the p-th BETTI group, H'_p, of V' ($p = 0, 1, \ldots, n$). The same application is supplied by all the correspondences which are homologous to T with division. Hence we obtain an essential representation, which, in virtue of (5), is itself linear, of the set of correspondences between V and V' into the set of linear applications between H_p and H'_p.

The inverse correspondences T^{-1}, between V' and V, are similarly associated with the applications between H'_p and H_p. Further, the process of passing to the inverse is a linear mapping between the direct and inverse correspondences, that is, with the obvious notation (§ 83, p. 167), we have

$$(c_1 T_1 + c_2 T_2 + \cdots + c_k T_k)^{-1} = c_1 T_1^{-1} + c_2 T_2^{-1} + \cdots + c_k T_k^{-1}.$$

It should be realized that T and T^{-1} do not always induce between H_p, H'_p and between H'_p, H_p linear applications which are mutually inverse; in fact this does not generally occur, as we see, for example, from equations (9).

We conclude this section by remarking that *the product of two correspondences, as we see by considering them as relations* (§ 82, p. 165), *is a relation, but in general no longer a correspondence.* This follows from the fact that, on a topological variety, the set of points common to two cycles does not itself necessarily consist of the points of a cycle. In fact, only by considering the homology classes of the two, can we introduce (not a single cycle but) a third homology class which can be defined as the "intersection" of the other two classes. On the other hand we note that, for the study of one-valued correspondences between two varieties, we must often regard the correspondences as cycles (and not simply as homology classes) on the product of the two varieties.

Thus, in the general theory of correspondences between topological varieties, it is not permissible — without suitable precautions — to introduce the products of two correspondences. Possibly it is for this reason that such products have seldom been considered in topology. We shall return to this question later (§§ 89—95, pp. 173—183), from a constructive point of view.

§ 85. Topological invariants of correspondences between topological varieties. — Retaining the terminology and notations of §§ 82—83, we recall that, with each correspondence T between V and V', there is associated a linear mapping of H_p into H'_p, and similarly, with the inverse transformation T^{-1}, there is associated a linear mapping of H'_p into H_p ($p = 0, 1, \ldots, n$). For simplicity, we shall denote these mappings by the same symbols, Θ_p and Θ'_p, as the matrices which enter in their representations (4) and (4'). Each of the mappings obtained by composition of Θ_p and Θ'_p, namely each of the products

$$\Theta_p \Theta'_p \Theta_p \Theta'_p \ldots \Theta_p \Theta'_p, \quad \Theta_p \Theta'_p \Theta_p \Theta'_p \ldots \Theta'_p \Theta_p,$$
$$\Theta'_p \Theta_p \Theta'_p \Theta_p \ldots \Theta'_p \Theta_p, \quad \Theta'_p \Theta_p \Theta'_p \Theta_p \ldots \Theta_p \Theta'_p,$$

is consequently a topological covariant of T: this also holds for all the entities which are defined by these products in an intrinsic manner, i. e., those which are independent of the bases of H_p and H'_p.

In this way, we obtain a whole set of *topological covariants* of T, which may be of an arithmetic, algebraic or geometric character.

We shall now illustrate these remarks; but, for greater clarity of exposition, we shall restrict ourselves to the linear applications defined by the symbols

$$\Theta_p, \quad \Theta'_p, \quad \Omega_p = \Theta_p \Theta'_p, \quad \Omega'_p = \Theta'_p \Theta_p ;$$

it should be noticed that the final two operate in H_p and in H'_p respectively.

§ 86. Arithmetic and algebraic invariants. — The simplest topological invariants of T are the integers (positive, zero or negative) α and α' introduced towards the end of §81, p. 165; these integers are called, respectively, the first and second *indices* of T. Recalling the results of §§ 82, 83, pp. 165—167, we see immediately that the indices of T^{-1} are α' and α. Further, with the notation of § 81, p. 164, it follows from (4), (4'), (5) and (5') that, for $p = 0$ and $p = n$, we have

$$TP \sim \alpha' P', \quad TV \sim \alpha V',$$
$$T^{-1} P' \sim \alpha P, \quad T^{-1} V' \sim \alpha' V .$$

These relations show that the first index α of T is equal both to the algebraic number of points of V which correspond in T^{-1} to a point P' of V', and to the algebraic number of coverings of the variety V' which is the transform of V by means of T. We have, of course, a similar significance of the number α'.

For the particular case in which T is one-valued, i. e. when T is a continuous application of V into V', we can choose the orientations such that $\alpha' = 1$; then α coincides with the degree of T (in the sense of Brouwer).

Other topological invariants of T, of an arithmetic or algebraic character, are the *ranks* of the matrices Θ_p, Θ'_p, Ω_p, Ω'_p; we shall denote these by

$$\varrho_p, \varrho'_p, \sigma_p, \sigma'_p$$

respectively; further, the *latent roots* of Ω_p and Ω'_p, which differ from zero, are also invariants. Relative to the first of these invariants, we deduce that, since the rank of the product of two matrices does not exceed the rank of either of its factors, the following chain of inequalities holds

$$(\sigma_p \text{ and } \sigma'_p) \leqq (\varrho_p \text{ and } \varrho'_p) \leqq (R_p \text{ and } R'_p) .$$

On the other hand, since A_p and A'_p are non-degenerate, we have from (5) and (5') the following equality

$$\varrho_p = \varrho'_{n-p} = \text{rank } (X^{n-p}) .$$

Whence we see that, *if Ω'_p is non-degenerate* (i. e. if Ω'_p is a mapping of H'_p onto itself), then

$$\sigma'_p = \varrho_p = \varrho'_p = R'_p \quad \text{and} \quad R'_p \leqq R_p .$$

An analogous result is obtained if Ω_p is non-degenerate; so that, *if* Ω_p *and* Ω'_p *are both non-degenerate*, we have

$$\sigma_p = \sigma'_p = \varrho_p = \varrho'_p = R_p = R'_p.$$

We can also show, finally, that *the non-zero latent roots of* Ω_p *coincide with the non-zero latent roots of* Ω'_p; moreover, *the latter have respectively the same multiplicities as the former* (cf. B. SEGRE [122], pp. 235—236).

§ 87. Geometric invariants. — We obtain other topological invariants of T, of a geometric character, by consideration of the first axes and the second axes of the linear mappings $\Theta_p, \Theta'_p, \Omega_p, \Omega'_p$ (defined at the end of § 85, p. 169). We denote these axes by

$$1_{\Theta_p}, 1_{\Theta'_p}, 1_{\Omega_p}, 1_{\Omega'_p} \quad \text{and} \quad 2_{\Theta_p}, 2_{\Theta'_p}, 2_{\Omega_p}, 2_{\Omega'_p}$$

respectively; where, for example, 1_{Θ_p} is the sub-module of H_p (with dimension $R_p - \varrho_p$) formed by the cycles, C_p, such that $TC_p \approx 0$; and 2_{Θ_p} is the sub-module of H'_p (with dimension ϱ_p) formed by the cycles, C'_p, satisfying relations of the form $C'_p \approx TC_p$ (for some C_p of V).

It is not difficult to establish the relations

$$1_{\Theta_p} \subseteq 1_{\Omega_p}, \; 2_{\Omega_p} \subseteq 2_{\Theta'_p}, \; 1_{\Theta'_p} \subseteq 1_{\Omega'_p}, \; 2_{\Omega'_p} \subseteq 2_{\Theta_p}.$$

We have, moreover,

$$\dim \left(1_{\Theta_p} \cap 2_{\Theta'_p}\right) = \varrho'_p - \sigma'_p \quad \text{and} \quad \dim \left(1_{\Theta'_p} \cap 2_{\Theta_p}\right) = \varrho_p - \sigma_p.$$

Obviously, the left-hand sides of these equations cannot exceed $\dim 1_{\Theta_p} = R_p - \varrho_p$ and $\dim 1_{\Theta'_p} = R'_p - \varrho'_p$. Whence we immediately deduce the inequalities

$$R_p \geqq \varrho_p + \varrho'_p - \sigma'_p \quad \text{and} \quad R'_p \geqq \varrho_p + \varrho'_p - \sigma_p.$$

For the case in which the mapping Ω'_p is *non-degenerate*, we know (§ 86, p. 169) that $\varrho_p = \varrho'_p = \sigma'_p$; consequently, then,

$$\dim \left(1_{\Theta_p} \cap 2_{\Theta'_p}\right) = 0,$$

and

$$\dim \left(1_{\Theta_p} \cup 2_{\Theta'_p}\right) = \dim 1_{\Theta_p} + \dim 2_{\Theta'_p} = (R_p - \varrho_p) + \varrho'_p = R_p.$$

Therefore H_p *is then the direct sum of* 1_{Θ_p} *and* $2_{\Theta'_p}$. This is equivalent to saying that each cycle C_p of V satisfies one, and only one, homology of the form

$$C_p \approx C_p^{(1)} + C_p^{(2)}, \quad \text{where} \quad TC_p^{(1)} \approx 0 \quad \text{and} \quad C_p^{(2)} \approx T^{-1} C'_p.$$

We note, finally, that, when Ω'_p is *totally degenerate*, i. e. when $\Omega'_p = 0$, we have $\sigma'_p = 0$. From which it follows that

$$\dim \left(1_{\Theta_p} \cap 2_{\Theta'_p}\right) = \dim 2_{\Theta'_p},$$

i. e.

$$2_{\Theta'_p} \subseteq 1_{\Theta_p};$$

and this leads to the relations

$$\dim 2_{\Theta'_p} \leqq \dim 1_{\Theta_p} \quad \text{or} \quad \varrho'_p \leqq R_p - \varrho_p .$$

But we also know in any case (§ 86, p. 169) that $\varrho'_p = \varrho_{n-p}$; so that, for the present case, we obtain

$$R_p \geqq \varrho_p + \varrho_{n-p} .$$

§ 88. *i*-correspondences on topological varieties. — Throughout this section, we shall suppose that the variety V' is superimposed on V, and that the two varieties have the same orientation. Then, on the product $W = V \times V'$, we can consider the "symmetry", S, which associates two points as $P \times P'$ and $P' \times P$; thus, the united points of S are the points of the diagonal variety Δ of W (§ 82, p. 166). It is immediately seen that S is a correspondence of W onto itself, in the sense of § 82, and that it is completely determined from this point of view by the condition

$$S\Delta \approx \Delta , \tag{10}$$

which has an obvious geometric significance.

On V we can fix a Betti base, consisting of p-cycles C^i_p ($i = 1, 2, \ldots, R_p$; $p = 0, 1, \ldots, n$), which reduces to a point if $p = 0$, and each of which is an irreducible circuit if $p > 0$ (we admit the existence of such a base, in order to simplify the argument which follows: but this admission could be avoided). Further, it can be arranged that none of the elements x^p_{ij} of the matrix (6) are zero. Then, by choosing $C'^i_p = C^i_p$, we see that $C^i_p \times C'^j_{n-p}$ is an irreducible n-circuit of W, and that this circuit is transformed by S into an n-cycle which can only differ by an integer factor, $\lambda^{ij}_{n,p}$, from the associated n-circuit $C^j_{n-p} \times C'^i_p$.

But we also know (§ 82, p. 166) that

$$\Delta \approx \sum_0^n {}_p\, x^p_{ij} (C^i_p \times C'^j_{n-p}) ,$$

where the matrix of coefficients, $X^p = \|x^p_{ij}\|$, is given by (6); and, in virtue of (1), we have the equations $x^p_{ij} = (-1)^{p(n-p)} x^{n-p}_{ji}$. Consequently, the homology (10) can be written in the form

$$0 \approx S\Delta - \Delta \approx \sum_0^n {}_p \{\lambda^{ij}_{n,p} - (-1)^{p(n-p)}\}\, x^p_{ij} (C^j_{n-p} \times C'^i_p) ,$$

which provides the equalities

$$\lambda^{ij}_{n,p} = (-1)^{p(n-p)} \quad (i, j = 1, 2, \ldots, R_p;\ p = 0, 1, \ldots, n).$$

We have thus established the homologies

$$S(C^i_p \times C'^j_{n-p}) = (-1)^{p(n-p)} (C^j_{n-p} \times C'^i_p) ,$$

which could have been proved in a more direct, but rather less simple, manner by considerations on the product $W \times W$. From these relations, we deduce immediately that, if (2) now denotes any n-cycle Γ_n of W, its transform $S\Gamma_n$ by means of S satisfies the homology

$$S\Gamma_n \approx \sum_0^n {}_p (-1)^{p(n-p)} x_{ji}^{n-p} (C_p^i \times C'^{j}_{n-p}). \tag{11}$$

Hence the relation $S^2\Gamma_n = S(S\Gamma_n) \approx \Gamma_n$ always holds, which is in accordance with the involutory character of S.

We shall say that a *correspondence* T of V into itself is *involutory* or, briefly, that it is an *i-correspondence*, if, whenever P, P' denotes any pair of points of V which, in this order, correspond in T, then P' and P also correspond in T. In this case, the cycle Γ_n which represents T on W (§ 82, p. 165) — considered as a point-set — is transformed into itself by S, and conversely. But this does not imply, as we shall shortly see (cf. the theorem below), that the homology $S\Gamma_n \approx \Gamma_n$ necessarily holds. However, it is clear that the i-correspondences of V into itself always constitute a *module.*

If the n-cycle, Γ_n, which represents an i-correspondence is an irreducible circuit of W, then there exists, manifestly, an integer λ such that $S\Gamma_n \approx \lambda\Gamma_n$. From the above results we now deduce that

$$\Gamma_n \approx S(S\Gamma_n) \approx S(\lambda\Gamma_n) \approx \lambda S\Gamma_n \approx \lambda^2\Gamma_n;$$

hence $\lambda^2 = 1$, i. e. $\lambda = 1$ or $\lambda = -1$, if $\Gamma_n \not\approx 0$.

An i-correspondence, T, will be called of the 1-st *species* or of the 2-nd *species* if, for its representative n-cycle Γ_n, we have respectively

$$S\Gamma_n \approx \Gamma_n, \tag{12}$$

or

$$S\Gamma_n \approx -\Gamma_n. \tag{13}$$

We shall shortly see that both these possibilities can in fact arise. However, from the foregoing, there certainly exist no further cases for an i-correspondence, when it is represented by an irreducible circuit. We deduce from equations (12) and (13) that:

The i-correspondences of the 1-*st species and the i-correspondences of the* 2-*nd species constitute modules. Two i-correspondences, one of the* 1-*st species and one of the* 2-*nd species, neither of which is homologous with division to zero, have as their sum an i-correspondence which is neither of the* 1-*st nor of the* 2-*nd species.*

We note that there remains the question of deciding if, or for which varieties V, it is impossible to decompose all its i-correspondences in this manner. If we now substitute the expression for Γ_n and $S\Gamma_n$, given by (2) and (11), into the relations (12) and (13), then, recalling (5), we deduce that:

An i-correspondence T (for which we use the notation of §§ 81, 82, *pp.* 164—166*) is of the* 1-*st species if, and only if,*

$$X^p = (-1)^{p(n-p)} (X^{n-p})_{-1} \qquad (p = 0, 1, \ldots, n) \quad (14)$$

(for $p = 0$, we then have $\alpha = \alpha'$), and this is equivalent to

$$\Theta_p = A_p (\Theta_{n-p})_{-1} (A_p)^{-1} \qquad (p = 0, 1, \ldots, n). \quad (15)$$

T is of the 2-*nd species if, and only if,*

$$X^p = -(-1)^{p(n-p)} (X^{n-p})_{-1} \qquad (p = 0, 1, \ldots, n) \quad (16)$$

(which, for $p = 0$, gives $\alpha = -\alpha'$), and this is equivalent to

$$\Theta_p = -A_p (\Theta_{n-p})_{-1} (A_p)^{-1} \qquad (p = 0, 1, \ldots, n). \quad (17)$$

We consider, for example, the i-correspondence of the n-sphere into itself which transforms each point into the diagonally opposite point. We see that it has indices $[(-1)^{n+1}, 1]$ and that it is of the 1-st species if n is odd and of the 2-nd species if n is even.

In particular, from (14) and (16), we deduce that, if $n = 2p$ is even, then the matrix X^p must be symmetric when the i-correspondence T is of the 1-st species, and skew-symmetric when T is of the 2-nd species. Moreover, in general, equations (15) imply that $\operatorname{tr} \Theta_p = \operatorname{tr} \Theta_{n-p}$; and, similarly, from (17) it follows that $\operatorname{tr} \Theta_p = -\operatorname{tr} \Theta_{n-p}$ $(p = 0, 1, \ldots, n)$. We now make use of the formula (8) of LEFSCHETZ, in order to obtain the following table, concerning the algebraic number $u = u(T)$ of united points of an i-correspondence of the 1-st or the 2-nd species.

	T of 1st *species*	*T of* 2nd *species*
n odd	$u = 0$	$u \equiv 0 \pmod 2$
$n = 2p$ even	$u \equiv \operatorname{tr} \Theta_p \pmod 2$	$u = (-1)^p \operatorname{tr} \Theta_p$

More particularly, if the BETTI numbers of the variety V are the same as those of an n-sphere, then the table can be made more precise, as follows:

	T of 1st *species*	*T of* 2nd *species*
n odd	$u = 0$	$u = 2\alpha'$
n even	$u = 2\alpha'$	$u = 0$

§ 89. Semiregular correspondences and their products. — We have already seen in § 84, p. 168, that, if we want the product of two correspondences between topological varieties to be again a correspondence, then we must suitably restrict the notion of a correspondence. In the sequel, the restrictions we impose will be found to be useful also for other considerations.

The two n-dimensional topological varieties V, V' considered, we shall suppose to be (not only connected, closed and orientable, but also) differentiable. We shall not enter here into questions concerning less restrictive conditions, which may be imposed on the varieties, in order that the properties in the sequel should still be valid (for details relative to such considerations, cf. B. SEGRE [122] n. 91, pp. 250—252).

Let P and P' be a pair of points of V and V' which are homologous in the relation T between the varieties. If $(x_1, x_2, \ldots, x_n)$ and $(x'_1, x'_2, \ldots, x'_n)$ denote allowable coordinates in the neighbourhoods of P and P' on V and V', then we shall say that T is *regular* in the points P, P' when it can be represented in the neighbourhoods of these points by equations of the form

$$x'_i = \varphi_i(x_1, x_2, \ldots, x_n) \qquad (i = 1, 2, \ldots, n), \tag{18}$$

where the functions φ_i are of class C^1 and the Jacobian of the transformation is non-zero. For this condition to be satisfied, it is necessary and sufficient that the image, Γ, of T on the product variety $W = V \times V'$ should have a simple point at $Q = P \times P'$, that it should be n-dimensional at Q, and that it should touch neither $P \times V'$ nor $V \times P'$ at Q.

Further, we shall say that a relation T between V and V' is semi-regular when the image Γ of T on W is a closed set, and when T is regular almost everywhere, i. e. regular at all homologous pairs of points apart from, at most, those pairs represented on Γ by the points of a subcomplex of W with dimension not exceeding $n - 2$. In this case we can show that, given a suitable orientation, Γ is an n-cycle of W; this orientation can, for example, be chosen such that, in the neighbourhood of every homologous pair P, P' at which T is regular, the positive orientations of Γ and V agree when Γ is projected onto V. Hence T has now become a correspondence (in the strict sense of § 82, p. 165). We shall say that such a correspondence T, oriented as suggested above, is an *elementary correspondence*. Every correspondence which is a linear combination — with integral coefficients — of elementary correspondences will be termed a *semi-regular correspondence*, or, more briefly, an *s-correspondence*.

It is seen immediately that *the inverse of a semi-regular correspondence is also semi-regular*. An s-correspondence, T, will be called effective when it is possible to choose the orientations such that T and T^{-1} are linear combinations with positive coefficients of elementary correspondences. We now introduce the notion of the product of s-correspondences with the following theorem.

The composition of the relation defined by any s-correspondence T between V and V', with that defined by any s-correspondence $\overline{T}$ between V' and V'', furnishes a relation between V and V''; with the latter is associated a well-defined s-correspondence between V and V'': this we shall call the *product,* $\bar{T}T$, *of* T *and* $\bar{T}$. *The linear application induced by* $\bar{T}T$ *between the* BETTI *groups* H_p, H_p'' *of* V, V'' (§ 82, *p.* 165), *is precisely the product of the linear applications induced by* T *and* $\bar{T}$ *between the* BETTI *groups of* V, V' *and* V', V'' *respectively.*

Using this theorem, together with the formula of LEFSCHETZ (§ 82, p. 166), we immediately deduce that:

For the case in which V'' *coincides with* V, *the s-correspondences* $\bar{T}T$ *and* $T\bar{T}$ *(transforming* V *and* V', *respectively, into themselves) have the same number of united points. This number can be regarded as the algebraic number of pairs of points of* V, V' *which correspond both in* T *and* $\bar{T}$. *This result is immediately extended to the cyclic products of any number of s-correspondences.*

§ 90. Characteristic integers of a semi-regular correspondence. — To each elementary correspondence T (§ 89, p. 174), between V and V', can be attached three integers λ, μ, ν, which we define as follows. Let P' be any point of V', chosen arbitrarily outside a certain subset of V' which has dimension not exceeding $n-2$; then there is a fixed number, ν, of points P of V, such that P, P' correspond in T. Amongst these points P, there is a certain number, λ, appearing with multiplicity $+1$ in the set $T^{-1}P'$, that can be constructed as was indicated in § 83, p. 167. There remains a number, μ, of points of the set, each of which is counted in $T^{-1}P'$ with multiplicity -1. It now follows that

$$\lambda + \mu = \nu, \quad \lambda - \mu = \alpha, \quad \text{so that } \nu \equiv \alpha \pmod 2,$$

where α is the 1-st index of T (§ 86, p. 169).

The definition of $\lambda = \lambda(T)$, $\mu = \mu(T)$ and $\nu = \nu(T)$ can be easily extended to any s-correspondence T, and then for example $\lambda(-T) = \mu(T)$. We shall also put

$$\lambda'(T) = \lambda(T^{-1}), \quad \mu'(T) = \mu(T^{-1}), \quad \nu'(T) = \nu(T^{-1}),$$

and we shall say that $\nu = \nu(T)$ and $\nu' = \nu'(T)$ are the 1-st and 2-nd *s-indices* of T.

It is not difficult to prove, with the hypotheses stated towards the end of the preceding section, that if (ν, ν') and $(\bar{\nu}, \bar{\nu}')$ are the s-indices of T and $\bar{T}$ respectively, then $(\nu\bar{\nu}, \nu'\bar{\nu}')$ *are the s-indices of the product* $\bar{T}T$. We conclude this section by remarking that, in virtue of the conventions adopted in § 89, p. 174, $\mu' = 0$ for every elementary correspondence so that $\nu' = \lambda' = \alpha'$ and *the 2-nd s-index coincides with the 2-nd index.*

§ 91. Involutory elementary s-correspondences. — We devote this section to the proof of the following theorem, and to the deduction of some corollaries.

Let T *be an elementary correspondence of a variety* V *into itself, which is involutory and has only a finite number of united points. Then* *each of these united points counts a positive number of times in the algebraic number, $u = u(T)$, of united points of T (defined as in* *§ 82, p. 166**).*

Firstly, we remark that it suffices to prove the theorem for the case in which each of the united points of T is simple, i. e. when each united point has multiplicity $+1$ or -1 in the set of united points. This follows from the fact that we can deform T homotopically, so that the resulting correspondence, T', is always an elementary i-correspondence; T' is in fact represented on the symmetric product variety $V \times V$ by an n-cycle, which is obtained from the image of T by a homotopic deformation. For further details regarding these observations, we refer the reader to B. SEGRE [122], n. 94, pp. 257—264. Thus, since T has only a finite number of united points, we can choose T', near to T, such that it has only a finite number of united points each of which is simple. Now, if the theorem is assumed for correspondences with simple united points, we see that each united point, O, of T is counted in the number u with a positive multiplicity; this multiplicity being precisely the number of united points of T' which tend to O as T' tends to T.

Suppose that O is a simple united point of T, and that $(x_1, x_2, \ldots, x_n)$ are a set of allowable coordinates on V in the neighbourhood of O, such that the x_i's all vanish at O. Then, in this neighbourhood, T can be represented by equations (18), where the φ_i's vanish at O and have continuous first derivatives. Since T is involutory, it follows that the equations can also be written in the form

$$x'_i = \sum_1^n {}_j a_{ij} x_j + \cdots \qquad (i = 1, 2, \ldots, n),$$

where the dots represent terms with infinitesimal orders in the x_j's greater than unity. Let I denote the unit matrix of order n, and let A be the matrix of the coefficients a_{ij}. As O is, by hypothesis, a simple united point of T, the matrix $I - A$ has a non-zero determinant, and the sign of this determinant is the multiplicity of O in the number u. But the involutory character of T implies that $A^2 = I$. It is now clear that it is possible to satisfy these conditions for A in only one way in the real number field, namely by taking $A = -I$. Whence we deduce

$$\det(I - A) = \det(2I) = 2^n > 0;$$

so that O is counted with multiplicity $+1$ in u, and the theorem is proved.

An immediate corollary of this theorem is:

The algebraic number, u, of united points of an elementary i-correspondence (which does not possess an infinity of united points) is always positive or zero. When this number is zero, the correspondence is free from united points.

We see that these results are in agreement with those summarized in the last table of § 88, p. 173, since, as we observed in § 90, p. 175, for an elementary correspondence we always have $\alpha' = \nu' > 0$.

Further, using the results summarized in the first table of § 88, p. 173, we deduce, from the above properties, that:

On a differentiable variety of odd dimension n, every elementary i-correspondence of the 1-*st species either has no united points or has an infinity of such points.*

On a differentiable variety of even dimension $n = 2p$, to each elementary i-correspondence of the 2-*nd species (which does not admit an infinity of united points) there is attached a matrix Θ_p* (*§ 82, p.* 166). *The trace of Θ_p is either zero (and this certainly occurs if $R_p = 0$), in which case the correspondence has no united points, or it has the sign $(-1)^p$, and then the correspondence is definitely not free from united points.*

§ 92. Algebraic and skew-algebraic involutory transformations. — The results of the preceding section, and those of the sections which follow, can be used in many ways: and particularly important are the applications to algebraic geometry. Considering any algebraic correspondence between two complex algebraic varieties of the same dimension, such that the correspondence too has this common dimension and is without degenerate components, we note that it gives rise to an s-correspondence between the Riemannians of the two varieties, even when it possesses exceptional varieties. If (as we shall suppose in the sequel) the algebraic correspondence is effective, then the associated s-correspondence is also effective (in the sense of § 89, p. 174). Properties similar to these also arise from skew-algebraic correspondences. Moreover, if F is an irreducible (complex) algebraic variety free from singularities and having dimension $m > 0$, then every *involutory algebraic* ∞^m *self-transformation* of F induces, on the Riemannian V (of dimension $n = 2m$) of F, an i-correspondence of the 1-st species. Similarly, the image on V of an *involutory skew-algebraic transformation* of F is an i-correspondence which is of the 1-st or 2-nd species, according as m is even or odd.

From the final theorem of § 91, p. 177, we can now deduce, for example, that:

If a complex algebraic variety F has odd dimension m, and if the m-th Betti *number of its Riemannian is zero, then every involutory skew-algebraic self-transformation of F is either without united points or contains an infinity of united points.*

This result is classical for the very special case in which F is a complex projective space with odd dimension, and the correspondence in question is supposed to be a skew-homography of the space (cf. C. Segre [125] nn. 18, 23).

§ 93. An extension of Zeuthen's formula to the topological domain. — Let T be an s-correspondence between two differentiable varieties V, V', for which we retain the notation of §§ 82, 83, 89, 90, pp. 165—175. If T or $-T$ is an elementary correspondence, then, for T, we have $\alpha = \pm \nu \neq 0$, $\alpha' = \pm \nu' \neq 0$. We now put

$$\alpha_1 = \alpha_1(T) = \alpha \operatorname{sign}(\alpha') \quad \text{and} \quad \alpha_1' = \alpha_1'(T) = \alpha' \operatorname{sign}(\alpha),$$

so that $\alpha_1(-T) = \alpha_1(T)$, $\alpha_1'(-T) = \alpha_1'(T)$. In general we have

$$T = \Sigma\, c_i T_i,$$

where the T_i's are elementary correspondences and the c_i are integers. We define

$$\alpha_1 = \alpha_1(T) = \Sigma\, |c_i|\, \alpha_1(T_i), \quad \alpha_1' = \alpha_1'(T) = \Sigma\, |c_i|\, \alpha_1'(T_i),$$

and we note that, from § 90, p. 175, these relations give rise to the equations

$$\nu(T) = \Sigma\, |c_i|\, \nu(T_i), \quad \nu'(T) = \Sigma\, |c_i|\, \nu'(T_i).$$

In particular, for an effective s-correspondence (§ 89, p. 174), we can, by a suitable choice of the orientations of V and V', always ensure that the relations

$$\alpha(T) = \alpha_1(T) = \nu(T), \quad \alpha'(T) = \alpha_1'(T) = \nu'(T)$$

hold.

We now proceed with the consideration of the product

$$U' = T\,T^{-1}.$$

First we note that U' is an s-correspondence of V' into itself, for which the two s-indices are equal to $\nu\nu'$. Amongst the $\nu\nu'$ points of V' which correspond in U' to a point P' generally situated on V', there is always the point P' itself, counted ν times. This point is, in fact, counted a certain number of times positively and a certain number of times negatively, and it can be seen that the algebraic number which is the multiplicity of P' is equal to $\alpha_1(T)$. This means that, in the present case, the identical correspondence I' of V' into itself is a part of U', and is indeed counted α_1 times.

The residual correspondence

$$L' = T\,T^{-1} - \alpha_1 I' \tag{19}$$

is an s-correspondence of V' into itself, for which a general point of V' is no longer self-corresponding. We deduce easily that L' is involutory according to the definition of § 88, p. 172. In fact, two (distinct) points of V' are associated in L' if, and only if, they correspond by means of T to the same point of V. We shall see later in this section that L' is always an i-correspondence of the 1-st species.

It is evident, from § 90, p. 175, that L' has both of its s-indices equal to $\nu(\nu'-1)$: but this is zero if $\nu'=1$. Hence $L'=0$ *when* T *is one-valued*, i. e. when $\nu'=1$; therefore, in this case,

$$T T^{-1} = \alpha_1 I'. \tag{20}$$

If we now exclude this case, we see that the united points of L', which are not exceptional for T', are precisely those points which count at least twice in the set of ν' points of V' corresponding in T to some point P of V. Such a point P of V is called a *branch-point* of T on V. Thus the algebraic number δ' of united points of L' can be defined as the *algebraic number of branchings of* T *on* V.

Similarly, the *algebraic number of branchings of* T *on* V' is simply the algebraic number δ of united points of the involutory s-correspondence L, of V into itself, where

$$L = T^{-1} T - \alpha_1' I. \tag{21}$$

We note that, in view of (19) and (21), the correspondences L, L' are unaltered if we replace T by $-T$. Moreover, if T or $-T$ is an elementary correspondence, the indices of which can be assumed to be positive, then each of L, L' reduces to a positive sum of elementary correspondences. Whence we can apply the theorem of § 91, p. 175, and deduce that

$$\delta \geqq 0, \quad \delta' \geqq 0.$$

An equality sign in these relations would imply that T (*on* V' *and* V *respectively*) *has either no branch-point or an infinity of such points.* Further, this result can be immediately extended to the case in which T is an *effective s-correspondence.*

We shall now *calculate the numbers* δ *and* δ'. We denote by Λ_p' the square matrix of order R_p' $(p = 0, 1, \ldots, n)$ which is attached to the correspondence L' in a manner similar to that by which the matrices Θ_p and Θ_p' are attached to T and T^{-1} (§§ 82, 83). It follows, from (19) and § 89, p. 174, that Λ_p' is given by

$$\Lambda_p' = \Theta_p' \, \Theta_p - \alpha_1 I_{R_p'} \qquad (p = 0, 1, \ldots, n), \tag{22}$$

where $I_{R_p'}$ is the unit matrix of order R_p'. The required expressions are now obtained from equations (22), when we utilize them with the formula of Lefschetz (8): thus

$$\delta' = \sum_0^n {}_p (-1)^p \operatorname{tr}(\Theta_p' \, \Theta_p) - \alpha_1 \chi', \tag{23}$$

where $\chi' = \sum_0^n {}_p (-1)^p R_p'$ is the Euler-Poincaré characteristic of V'.

The matrices Θ_p, Θ_p' appearing in (22) can be replaced by the expressions given in (5) and (5′), so that we obtain

$$\Lambda_p' = (-1)^{p(n-p)} A_p' (X^p)_{-1} A_p X^{n-p} - \alpha_1 I_{R_p'}. \tag{24}$$

Similarly, (23) could be written in an alternative form. Moreover, from (22) and (9) we deduce that

$$\Lambda'_p = A'_p (\Lambda'_{n-p})_{-1} (A'_p)^{-1} \qquad (p = 0, 1, \ldots, n).$$

These equalities then give rise to the following results, when we recall the properties established in § 88, p. 173.

The correspondence L', defined on V' as above (and similarly the correspondence L on V), is always an i-correspondence of the 1-*st species.*

By interchanging the roles of V and V', we obtain the analogue of (23):

$$\delta = \sum_0^n {}_p (-1)^p \operatorname{tr}(\Theta_p \Theta'_p) - \alpha'_1 \chi. \tag{25}$$

Noting that $\operatorname{tr}(\Theta_p \Theta'_p) = \operatorname{tr}(\Theta'_p \Theta_p)$, we subtract (23) from (25) and so derive the important equation

$$\delta - \delta' = \alpha_1 \chi' - \alpha'_1 \chi. \tag{26}$$

This relation generalizes the classical formula of Zeuthen:

$$\delta - \delta' = 2\nu'(p-1) - 2\nu(p'-1),$$

relative to the (extremely particular) case in which T is an algebraic correspondence between the Riemannians of two algebraic curves of genera p and p'.

We remark that, if n is odd, in virtue of § 88, $\delta = \delta' = 0$. Hence, *if n is odd, every correspondence T which satisfies the conditions stated immediately after* (21) *either possesses no branch point or an infinity of such points.*

Similarly, if n is even and T satisfies those same conditions, the sum which enters into (23) or (25) cannot be less than $\alpha'_1 \chi$ or $\alpha_1 \chi'$ respectively.

Finally, we note that, if n is even and

$$R_p R'_p = 0 \qquad (p = 1, \ldots, n-1),$$

which occurs, for example, when one of V, V' is an n-sphere, then equations (23), (25) reduce to the much simpler relations

$$\delta = 2\alpha\alpha' - \alpha'_1 \chi, \qquad \delta' = 2\alpha\alpha' - \alpha_1 \chi'. \tag{27}$$

§ 94. One-valued elementary correspondences. — We shall now examine a little more deeply the case in which T is an *one-valued elementary correspondence.* With this hypothesis, we can choose the orientations such that

$$\alpha' = \nu' = 1, \qquad \alpha = \alpha_1 > 0, \qquad \alpha'_1 = 1. \tag{28}$$

Then, from (20), we deduce that

$$\delta' = 0\,, \tag{29}$$

$$\Theta'_p\,\Theta_p = \alpha\, I_{R'_p} \qquad (p = 0, 1, \ldots, n)\,. \tag{30}$$

If we now suppose that V and V' satisfy the hypotheses stated at the end of § 91, p. 176, we can show, by using the results of that section, that (27)—(29) lead to

$$\delta = 2\alpha - \chi \geqq 0, \qquad \chi' = 2\,.$$

Without these restrictions on V and V', we see, from (26), (28), (29) and the results of § 91, that

$$\delta = \alpha\,\chi' - \chi \geqq 0\,. \tag{31}$$

Whence we deduce that *there exists a one-valued elementary correspondence of a variety V, with characteristic χ, onto a variety V', with characteristic $\chi' \leqq 0$, only if $\chi \leqq \chi'$.*

With the notation introduced in § 85, p. 169, equations (30) become $\Omega'_p = \alpha\, I_{R'_p}$; so that, in the present case, none of the matrices Ω'_p can be degenerate. Therefore we can apply the results of §§ 86, 87, pp. 169, 170, which depend on this condition. For the sake of brevity, we shall only enunciate some of the results so obtained. We have in fact:

When a differentiable variety is subjected to a one-valued elementary correspondence, none of the Betti *numbers are augmented.*

Further we note that, if $R_p = R'_p \neq 0$, for a given value of p, then the corresponding equation of (30) takes the form

$$\Theta_p\,\Theta'_p = \alpha\, I_{R_p}\,.$$

From equation (19), we now see that the above relation is equivalent to

$$\Theta_p\, A'_p\,(\Theta_{n-p})_{-1} = \alpha\, A_p\,.$$

And, in virtue of §§ 81—82, pp. 164—176, this is merely a compact form of the equations

$$[T\,C^i_p, T\,C^{i'}_{n-p}] = \alpha\,[C^i_p, C^{i'}_{n-p}]\,.$$

Hence, for any p-cycle C_p and any $(n-p)$-cycle C_{n-p} of V, we have the equality

$$[T\,C_p,\, T\,C_{n-p}] = \alpha\,[C_p,\, C_{n-p}]\,,$$

which has an obvious geometric interpretation.

§ 95. Correspondences represented by differentiable varieties. — We conclude this final Part, by giving some properties of an s-correspondence T between differentiable varieties V, V', when the indices of T take any values, but when its representative cycle Γ on $V \times V'$ is an orientable, irreducible differentiable variety. This then implies that either T or $-T$ is an elementary correspondence.

We can always choose the orientations of the varieties such that T has indices $\alpha = \alpha_1 > 0$ and $\alpha' = \alpha'_1 > 0$. Thus the projection of Γ onto each of V and V' is a one-valued elementary correspondence. We shall call these correspondences $\bar{T}$ and $\bar{T}'$. It is clear that they have indices $(\alpha', 1)$ and $(\alpha, 1)$. Let $\bar{\delta}'$ and $\bar{\delta}$ be the algebraic numbers of branch-points of $\bar{T}$ on V and of $\bar{T}'$ on V' respectively; then $\bar{\delta}'$ and $\bar{\delta}$ are simply the algebraic numbers of united points of the self-correspondences of Γ defined by the equations

$$\bar{L}' = \bar{T}^{-1}\bar{T} - \bar{I} \quad \text{and} \quad \bar{L} = \bar{T}'^{-1}\bar{T}' - \bar{I} \tag{32}$$

respectively, where $\bar{I}$ is the identity correspondence of Γ into itself (§ 93, p. 178). Moreover, if $\bar{\chi}$ is the EULER-POINCARÉ characteristic of Γ, then, by applying the formulae (20) and (31) to the correspondences $\bar{T}$ and $\bar{T}'$, we deduce that

$$\bar{T}\,\bar{T}^{-1} = \alpha' I\,, \quad \bar{T}'\bar{T}'^{-1} = \alpha I'\,, \tag{33}$$

$$\bar{\delta}' = \alpha'\chi - \bar{\chi} \geqq 0\,, \quad \bar{\delta} = \alpha\chi' - \bar{\chi} \geqq 0\,. \tag{34}$$

Hence

$$\bar{\delta} - \bar{\delta}' = \alpha\chi' - \alpha'\chi\,. \tag{35}$$

On the other hand, we easily establish that

$$T = \bar{T}'\bar{T}^{-1} \quad \text{and, therefore,} \quad T^{-1} = \bar{T}\,\bar{T}'^{-1}\,.$$

The relations (19) and (21) now become

$$L' = \bar{T}'\,\bar{T}^{-1}\,\bar{T}\,\bar{T}'^{-1} - \alpha I' \quad \text{and} \quad L = \bar{T}\,\bar{T}'^{-1}\,\bar{T}'\,\bar{T}^{-1} - \alpha' I\,.$$

Thus, with the aid of (32) and (33), we can write these equalities in the form

$$L' = \bar{T}'\,\bar{L}'\,\bar{T}'^{-1} \quad \text{and} \quad L = \bar{T}\,\bar{L}\,\bar{T}^{-1}. \tag{36}$$

In general (as in § 82, p. 166), we denote by $u(U)$ the algebraic number of united points of a correspondence U between two superimposed varieties. By application of the final proposition of § 89, p. 175, we deduce from (32) and (36):

$$u(L) = u(\bar{T}^{-1}\bar{T}\,\bar{L}) = u(\bar{L}'\bar{L} + \bar{L}) = u(\bar{L}'\bar{L}) + u(\bar{L})\,.$$

But, by definition, we have $u(L) = \delta$, $u(\bar{L}) = \bar{\delta}$. Thus, if we also put

$$\gamma = u(\bar{L}\bar{L}') = u(\bar{L}'\bar{L})\,,$$

we obtain

$$\delta = \bar{\delta} + \gamma \quad \text{and, similarly,} \quad \delta' = \bar{\delta}' + \gamma. \tag{37}$$

Whence it follows that

$$\delta - \delta' = \bar{\delta} - \bar{\delta}';$$

and we may notice that this equality is also an immediate consequence of (26) and (35). Finally, formulae (25), (34) and (37) lead to

$$\sum_0^n {}_p (-1)^p \operatorname{tr}(\Theta_p \Theta_p') = \bar{\delta} + \bar{\delta}' + \bar{\chi} + \gamma,$$

which expresses the sum appearing in (23) and (25) in terms of the characters introduced afterwards.

The foregoing results become more meaningful, and considerably simplified, for the case in which $\gamma = 0$. This occurs, for example, when the correspondence $\bar{L}\bar{L}'$ has no united elements. It should be noted that a united element of $\bar{L}\bar{L}'$ is simply a pair of points of V, V' such that each one is counted at least twice in the set of points which are the transforms of the other by means of T or T^{-1}.

Historical Notes and Bibliography

The results of this final Part have all been previously given in B. Segre [118], and also in B. Segre [122] (Chap. II, §§ 16—17, pp. 224—272). In this latter work the proofs which here have been only outlined are expounded in far greater detail.

The product correspondences of the type TT^{-1}, when T is a one-valued correspondence (here referred to in § 84), appeared for the first time in Hopf [55]. However, the results given there (and reproduced in Lefschetz [67] pp. 278—280) cannot be accepted without some reservations or further explanation, as is implied by our § 84, p. 168.

The i-correspondences appear rather incidentally in the literature, and with a much less general meaning than that given in § 88, p. 171; cf. for example the work of Smith [147], [148] (where, are to be found many results connected with periodic transformations, together with the relevant bibliography), and also that of Richardson [86], which concern the combinatorial aspect of the involutions generated, on a complex K, by a finite group (of order 2, and then of order > 2) of homomorphisms transforming each cell of K into a cell of K.

The formula of Zeuthen, of which we have given a generalization in § 93, p. 180, was given first in Zeuthen [172] (cf. also, for example, Severi [138]). This result of Zeuthen has been extended by Dedecker [41] to correspondences between real surfaces (one-sided or two-sided, open or closed). Other extensions to algebraic correspondences between algebraic varieties of dimensions 2, 3, and in general, have been successively given by Severi [134], Tafani [154], B. Segre [100], and Bassi [12].

With regard to some of the results of § 94, p. 181, we refer the reader to the more general results appearing in the paper of Hopf [55]. In this paper there are no considerations relative to the branch-points of correspondences, and, moreover, some of the signs appearing in the formulae of that work differ from those of the present treatment. However, regarding this treatment, attention should be paid to the earlier warning.

Bibliography

[1] AGOSTINELLI, C.: Sul prodotto di due determinanti con elementi complessi coniugati. Per. di Mat. (4) **13**, 252—254 (1933).

[2] ALEXANDROFF, P., u. H. HOPF: Topologie. Vol. I. Berlin: Julius Springer 1935.

[3] ATIYAH, M. F.: A Lefschetz fixed point formula for elliptic differential operators, Simposio Internazionale di Geometria Algebrica (1965), Cremonese, Roma, 38–39, 1968.

[4] — Bott periodicity and the index of elliptic operators, Quart. J. Math. Oxford Ser. (2), **19**, 113—140 (1968).

[5] — The index of elliptic operators III, Ann. of Math., **87**, 546—604 (1968).

[6] — and R. BOTT: A Lefschetz fixed point formula for elliptic differential operators, Bull. Amer. Math. Soc. **72**, 245—250 (1966).

[7] — — Harvard Seminar 1965 (mimeographed).

[8] — — A Lefschetz fixed point formula for elliptic complexes I, Ann. of Math., **86**, 374—407 (1967).

[9] — — A Lefschetz fixed point formula for elliptic complexes II: Applications, Ann. of Math., **88**, 451—491 (1968).

[10] — and G. B. SEGAL: The index of elliptic operators II, Ann. of Math., **87**, 531—545 (1968).

[11] — and I. M. SINGER: The index of elliptic operators I, Ann. of Math., **87**, 484—530 (1968).

[12] BASSI, A.: Su di alcune formole di geometria delle varietà algebriche, Rend. Circ. Mat. Palermo **60**, 107—123 (1936).

[13] BERTINI, E.: Introduzione alla geometria proiettiva degli iperspazi. 2ª ed. Messina: Principato 1923.

[14] BLASCHKE, W.: Vorlesungen über Integralgeometrie. 2. Aufl. Leipzig u. Berlin: B. G. Teubner 1936.

[15] BOCHNER, S.: Compact groups of differentiable transformations. Ann. of Math. **46**, 372—381 (1945).

[16] BOL, G.: Projektive Differentialgeometrie Teil I (1950), Teil II (1954), Teil III. Göttingen: Vandenhoeck and Ruprecht 1967.

[17] BOMPIANI, E.: Sopra alcune estensioni dei teoremi di Meusnier e di Eulero. Atti Accad. Sci. Torino **48**, 279—296 (1912).

[18] — Determinazioni proiettivo-differenziali relative ad una superficie dello spazio ordinario. Atti Accad. Sci. Torino **59**, 203—223 (1924).

[19] — Sulle trasformazioni puntuali e di contatto nel piano. Rend. Accad. Lincei (6) **4**, 435—437 (1926)$_2$.

[20] — Sul contatto di due superficie. Rend. Accad. Lincei (6) **15**, 116—121 (1932).

[21] — Invarianti proiettivi di una particolare coppia di elementi superficiali del second'ordine. Boll. Un. Mat. Ital. (1) **14**, 237—243 (1935).

[22] — Un système de courbes invariantes par projectivités. C. r. Acad. Sci. (Paris) **201**, 1006—1008 (1935).

[23] — Sulla normalizzazione delle equazioni differenziali lineari. Rend. Accad. Lincei (6) **23**, 807—812 (1936)$_1$.

[24] — Forme normali delle equazioni differenziali lineari e loro signicato geometrico. Ann. Sci. Univ. Jassy **23**, 75—105 (1937).

[25] — Determinazione delle curve algebriche sghembe appartenenti a quadriche. Rend. Accad. Lincei (6) **29**, 229—234 (1939)$_1$.

[26] — Über zwei Kalotten einer Hyperquadric. Jber. dtsch. Math.-Ver. **49**, 143—145 (1939).

[27] Bompiani, E.: Calotte a centri allineati di superficie algebriche. Rend. Accad. Italia (7) **1**, 93–101 (1939).

[28] — Invarianti proiettivi e topologici di calotte di superficie e d'ipersuperficie tangenti in un punto. Rend. di Mat. (5) **2**, 261—291 (1941).

[29] — Invarianti proiettivi di calotte. Rend. Accad. Italia (7) **2**, 888—895 (1941).

[30] — Sul birapporto di quattro punti di una curva. Boll. Un. Mat. Ital. (2) **4**, 84—86 (1942).

[31] — Geometria proiettiva di elementi differenziali. Ann. di Mat. (4) **22**, 1—32 (1943).

[32] — Sugli invarianti proiettivi di due calotte superficiali. Atti Accad. Sci. Torino **80**, 184—190 (1944/45).

[33] — Elementi differenziali regolari e non regolari nel piano e loro applicazioni alle curve algebriche piane. Rend. di Mat. (5) **5**, 1—46 (1946).

[34] Borsuk, K.: Drei Sätze über die n-dimensionale Euklidische Sphäre. Fund. Math. **20**, 177—190 (1933).

[35] Buzano, P.: Invariante proiettivo di una particolare coppia di elementi di superficie. Boll. Un. Mat. Ital. (1) **14**, 93—98 (1935).

[36] — Invarianti proiettivi di una coppia di elementi superficiali del 2° ordine. Rend. di Mat. (5) **1**, 139—162 (1940).

[37] Cartan, E.: Sur les développantes d'une surface réglée. Bull. Acad. Roumaine **14**, 167—174 (1931).

[38] Cartan, H.: Les fonctions de deux variables complexes et le problème de la représentation analytique. J. de Math. (9) **10**, 1—114 (1931).

[39] Cigala, A. R.: Sopra un criterio di instabilità. Ann. di Mat. (3) **11**, 67—78 (1905).

[40] Darboux, G.: Leçons sur la théorie générale des surfaces, t. I, 2ème ed. Paris: Gauthier-Villars 1914.

[41] Dedecker, P.: Sur la notion d'involution et la formule de Zeuthen (2 notes). Bull. Acad. Roy. Belgique **29**, 680—687 (1943); **30**, 58—66 (1944).

[42] Del Pezzo, E.: Sugli spazi tangenti ad una superficie o ad una varietà immersa in uno spazio di più dimensioni. Rend. Accad. Sci. Napoli 1886, 3—7.

[43] Drude, P.: Ein Satz aus der Determinantentheorie. Götting. Nachr., 1887, 118—122.

[44] End, W.: Algebraische Untersuchungen über Flächen mit gemeinschaftlicher Curve. Math. Ann. **35**, 82—90 (1889).

[45] Enriques, F. e O. Chisini: Lezioni sulla teoria geometrica delle equazioni e delle funzioni algebriche. Vol. III. Bologna: Zanichelli 1924.

[46] Fatou, P.: Substitutions analytiques et équations fonctionelles à deux variables. Ann. Ecol. Norm. sup. (3) **41**, 67—142 (1924).

[47] Fichera, G.: Sull'esistenza delle funzioni potenziali nei problemi della fisica matematica. Rend. Accad. Lincei (8) **2**, 527—532 (1947).

[48] Fubini, G. e E. Čech: Geometria proiettiva differenziale. Vol. I and II. Bologna: Zanichelli 1926—27.

[49] Godeaux, L.: Mémoire sur les surfaces multiples. Mém. Acad. Roy. Belgique 1952, 1—80.

[50] — Les singularités des points de diramation isolés des surfaces multiples. 2nd Coll. Géom. Alg. (C.B.R.M.), Liège 1952, 225—241.

[51] — Recherches sur les points de diramation de troisième catégorie d'une surface multiple. Bull. Acad. Roy. Belgique 1953, 1013—1023, 1087—1093; 1954, 81—86, 200—208, 355—370.

[52] — Sur l'existence de surfaces multiples possédant des points de diramation de structure donnée. Rend. di Mat. (5) **14**, 42—47 (1954).

[53] GRÉVY, A.: Etude sur les équations fonctionelles. Ann. Sci. Ecol. Norm. sup. (3) 11, 249—323 (1894).

[54] HSIUNG, C. C.: Projective invariants of contact of two curves in space of n dimensions. Quart. J. Math. (2) **17**, 39—45 (1946).

[55] HOPF, H.: On some properties of one-valued transformations of manifolds. Proc. Nat. Acad Sci. **14**, 206—214 (1928).

[56] HUMBERT, G.: Application géométrique d'un théorème de JACOBI, J. de Math. (4) **1**, 347—356 (1885).

[57] JORDAN, C.: Cours d'Analyse de l'Ecole Polytechnique. T. III, 3ème ed. Paris: Gauthier-Villars 1915.

[58] KOENIGS, G.: Recherches sur les substitutions uniformes. Bull. Sci. Math. (2) **7**, 340—357 (1883).

[59] — Recherches sur les intégrales de certaines équations fonctionelles. Ann. Sci. Ecol. Norm. sup. (3) **1**, Suppl. 3—41 (1884).

[60] KÖNIG, J.: Einleitung in die allgemeine Theorie der algebraischen Größen. Leipzig: B. G. Teubner 1903.

[61] LASKER, E.: Zur Theorie der Moduln und Ideale. Math. Ann. **60**, 20—116 (1905).

[62] LATTÈS, S.: Sur les équations fonctionelles qui définissent une courbe ou une surface invariante par une transformation. Ann. di Mat. (3) **13**, 1—138 (1907).

[63] — Sur les transformations de contact. C. r. Acad. Sci. (Paris) **148**, 902—907 (1909).

[64] — Sur les formes réduites des transformations ponctuelles à deux variables. C. r. Acad. Sci. (Paris) **152**, 1566—1569 (1911).

[65] — Sur les formes réduites des transformations ponctuelles dans le domaine d'un point double. Bull. Soc. Math. France **39**, 309—345 (1911).

[66] LEAU, L.: Etude sur les équations fonctionelles à une ou à plusieurs variables. Ann. de Toulouse **11**, 1—100 (1897).

[67] LEFSCHETZ, S.: Topology. Amer. Math. Soc. Coll. Publ. Vol. **12**. New York 1930.

[68] LÉMERAY, M.: Sur les équations fonctionelles qui charactérisent les opérations associatives et les opérations distributives. Bull. Soc. Math. France **27**, 130—137 (1899).

[69] — Applications des fonctions doublement périodiques à la solution d'un problème d'itération. Bull. Soc. Math. France **27**, 282—285 (1899).

[70] LEVI, B., L. A. SANTALO y C. DE MARIA: Estudios numerativos sobre las variedades de contacto de las superficies en un espacio de n dimensions. Inst. Mat. Univ. Litoral **8**, 3—72 (1946).

[71] LEVI-CIVITA, T.: Sopra alcuni criteri di instabilità. Ann. di Mat. (3) **5**, 221—307 (1901).

[72] LONGO, C.: Studio numerativo sopra le varietà di contatto delle superficie in uno spazio a n dimensioni. Ann. Sc. Norm. sup. Pisa (3) **4**, 223—230 (1950).

[73] — Sul modello minimo degli elementi cuspidali del piano. Ann. Sc. Norm. sup. Pisa (3) **9**, 45—63 (1955).

[74] MANARA, C. F.: Approssimazione delle trasformazioni puntuali regolari mediante trasformazioni cremoniane. Rend. Accad. Lincei (8) **8**, 103—108 $(1950)_1$.

[75] MARCHIONNA, E.: Sopra una proprietà caratteristica delle curve algebriche appartenti ad una quadrica. Rend. Accad. Lincei (8) **16**, 205—209 $(1954)_1$.

[76] MARLETTA, G.: Rapporto anarmonico di quattro punti considerati su una data curva. Boll. Accad Gioenia di Catania (3) **17**, 1—3 (1941).

[77] MASCALCHI, M.: Un nuovo invariante proiettivo di contatto di due superficie. Boll. Un. Mat. Ital. (1) **13**, 45—49 (1934).

[78] MAYER, A.: Über unbeschränkt integrable Systeme von linearen totalen Differentialgleichungen und die simultane Integration linearer partieller Differentialgleichungen. Math. Ann. **5**, 448—470 (1872).

[79] MEHMKE, R.: Einige Sätze über die räumliche Collineation und Affinität, welche sich auf die Krümmung von Kurven und Flächen beziehen. Schlömilchs Z. **36**, 56—60 (1891).

[80] — Über zwei die Krümmung von Kurven und das Gauss'sche Krümmungsmaß von Flächen betreffende charakteristische Eigenschaften der linearen Punkttransformationen. Schlömilchs Z. **36**, 206—213 (1891).

[81] NETTO, E.: Vorlesungen über Algebra. Vol. II. Leipzig: B. G. Teubner 1900.

[82] PERRIN, R.: Sur la relation qui existe entre p fonctions entières de $p-1$ variables. C. r. Acad. Sci. (Paris) **106**, 1789—1791 (1888).

[83] PICONE, M. e G. FICHERA: Trattato di Analisi matematica. Vol. II. Roma: Tumminelli 1956.

[84] — e A. GHIZZETTI: Integrazione dei sistemi degeneri di equazioni differenziali ordinarie, a coefficienti costanti. Rend. Accad. Naz. Lincei (8) **19**, 195—199 (1955)$_2$.

[85] POINCARÉ, H.: Sur les courbes définies par les équations différentielles (quatrième partie). J. de Math. (4) **2**, 151—217 (1886).

[86] RICHARDSON, M.: Homology characters of symmetric products. Duke Math. J. **1**, 50—69 (1935).

[87] RITT, J. F.: Differential equations from the algebraic standpoint. Amer. Math. Soc. Coll. Publ. Vol. **14**. New York 1932.

[88] — Differential algebra. Amer. Math. Soc. Coll. Publ. Vol. **33**. New York 1950.

[89] SCHEREIER, O. u. E. SPERNER: Einführung in die analytische Geometrie und Algebra. Vol. II. Leipzig: B. G. Teubner 1935.

[90] SCORZA-DRAGONI, G.: Sulle funzioni olomorfe di una variabile bicomplessa. Mem. Accad. Italia **5**, 597—665 (1934).

[91] SEGRE, B.: I sistemi semplicemente infiniti di superficie (in particolare piani e sfere) e le loro traiettorie ortogonali. Atti Accad. Sci. Torino **61**, 172—179 (1925/26).

[92] — Sui sistemi continui di curve piane con tacnodo. Rend. Accad. Naz. Lincei (6) **9**, 970—974 (1929)$_1$.

[93] — Intorno alla teoria delle superficie proiettivamente deformabili ed alle equazioni differenziali ad esse collegate. Mem. Accad. Italia (1) **2**, 49—189 (1931).

[94] — Intorno alla parità di alcuni caratteri di una varietà algebrica di dimensione dispari. Boll. Un. Mat. Ital. (1) **13**, 93—95 (1934).

[95] — I birapporti sulle superficie non sviluppabili dello spazio, e le condizioni geometriche per l'equivalenza proiettiva fra queste (2 Note). Rend. Accad. Naz. Lincei (6) **21**, 656—660; 692—697 (1935)$_1$.

[96] — Sugli elementi curvilinei che hanno comuni le origini ed i relativi spazi osculatori. Rend. Accad. Naz. Lincei (6) **22**, 392—399 (1935)$_2$.

[97] — Le linee proiettive ed un invariante d'immersione di una curva su di una superficie. Rend. Accad. Naz. Lincei (6) **22**, 400—405 (1935)$_2$.

[98] — Invarianti differenziali relativi alle trasformazioni puntuali e dualistiche fra due spazi euclidei. Rend. Circ. Mat. Palermo **60**, 224—232 (1936).

[99] — Sulle varietà di Veronese a due indici (2 Note). Rend. Accad. Naz. Lincei (6) **23**, 303—309; 391—397 (1936)$_1$.

[100] — Quelques résultats nouveaux dans la géométrie sur une V_3 algébrique (Mém. couronné), Mém. Acad. Roy. Belgique (2) **14**, 3—99 (1936).

[101] SEGRE, B.: Un teorema sopra le superficie algebriche con due fasci unisecantisi, ed una relazione fra gli angoli sotto cui si incontrano due curve algebriche tracciate su di una sfera. Boll. Un. Mat. Ital. (1) **15**, 169—172 (1936).

[102] — Invarianti topologici relativi ai punti uniti delle trasformazioni regolari fra varietà sovrapposte. Rend Accad. Naz. Lincei (6) **24**, 195—200 $(1936)_2$.

[103] — Una proprietà di certi spazi di VEBLEN ed alcune estensioni al campo differenziale della nozione di birapporto. Scritti-Mat. offerti a Luigi Berzolari, Pavia, Rossetti **1936**, 5—26.

[104] — Un complemento al principio di corrispondenza per le corrispondenze a valenza zero sulle curve algebriche. Rend. Accad. Naz. Lincei (6) **29**, 201—205 $(1936)_2$.

[105] — Complemento al principio di corrispondenza per le corrispondenze a valenza, con punti uniti di molteplicità qualsiansi. Rend. Accad. Naz. Lincei (6) **29**, 250—257 $(1936)_2$.

[106] — Sui sistemi di equazioni differenziali a derivate parziali d'ordine qualunque, con una sola funzione incognita, lineari a coefficienti costanti. Rend. Accad. Naz. Lincei (6) **27**, 208—212 $(1938)_1$.

[107] — Sull'estensione della formula integrale di CAUCHY e sui residui degli integrali n-pli, nella teoria delle funzioni di n variabili complesse. Atti 1° Congr. Un. Mat. Ital. (Firenze 1937), p. 174—180. Bologna: Zanichelli 1938.

[108] — Sui residui relativi ai punti uniti delle corrispondenze fra varietà sovrapposte. Atti 1° Congr. Un. Mat. Ital. (Firenze 1937), p. 259—263. Bologna: Zanichelli 1938.

[109] — On tac-invariants of two curves in a projective space. Quart. J. Math. (2) **17**, 35—38 (1946).

[110] — Un'estensione delle varietà di Veronese ed un principio di dualità per forme algebriche (2 Note). Rend. Accad. Naz. Lincei (8) **1**, 313—318; 559—563 $(1946)_1$.

[111] — Caratterizzazione geometrica degli integrali abeliani e dei loro residui (2 Note). Rend. Accad. Naz. Lincei (8) **3**, 167—172; 172—179 $(1947)_2$.

[112] — Sul contatto di due varietà. Boll. Un. Mat. Ital. (3) **1**, 12—16 (1947).

[113] — Sui teoremi di BÉZOUT, JACOBI e REISS. Ann. di Mat. (4) **26**, 1—26 (1947).

[114] — Un nuovo metodo per lo scioglimento delle singolarità. Rend. Accad. Naz. Lincei (8) **3**, 411—414 $(1947)_2$.

[115] — Corrispondenze analitiche e trasformazioni cremoniane. Ann. di Mat. (4) **28**, 107—139 (1949).

[116] — Sulla perfezione delle coincidenze isolate, Rend. Accad. Naz. Lincei (8) **10**, 335—336 $(1951)_1$.

[117] — Una proprietà caratteristica in grande delle curve giacenti su di una quadrica. Rend. Accad. Naz. Lincei (8) **12**, 374—378 $(1952)_1$.

[118] — Recouvrements de sphères et correspondances entre variétés topologiques. Coll. sur les questions de réalité en géométrie. (C. B. R. M.) Liège **1955**, 149—175.

[119] — Sur l'algébricité des courbes ayant un ordre relatif réel convenable. J. de Math. **35**, 43—54 (1956).

[120] — Invarianti topologico-differenziali, varietà di Veronese e moduli di forme algebriche. Ann. di Mat. (4) **41**, 113—138 (1956).

[121] — Sui sistemi di equazioni differenziali lineari a coefficienti costanti (3 Note). Rend. Accad. Naz. Lincei (8), **20**, 271—277; 395—403; 531—539 $(1956)_1$.

[122] — Forme differenziali e loro integrali. Vol. II. Roma: Docet 1956.

[123] Segre, B.: Differentiable varieties embedded in a projective space and homogeneous systems of linear partial differential equations, Tensor (N.S.), 13, 101—110 (1963).

[124] — Geometria proiettivo-differenziale delle equazioni di Laplace di tipo parabolico, Convegno Internazionale di Geometria Differenziale (1967). Bologna: Zanichelli 1968.

[125] Segre, C.: Un nuovo campo di ricerche geometriche (Nota II). Atti Accad. Sci. Torino **25**, 430—457 (1889/90).

[126] — Su alcuni punti singolari delle curve algebriche, e sulla linea parabolica di una superficie. Rend. Accad. Naz. Lincei (5) **6**, 168—175 $(1897)_2$.

[127] — Su una classe di superficie degli iperspazi legate colle equazioni lineari alle derivate parziali di 2° ordine. Atti Accad. Sci. Torino **42**, 559—591 (1906/07).

[128] — Preliminari di una teoria delle varietà luoghi di spazi. Rend. Circ. Mat. Palermo **30**, 87—121 $(1910)_2$.

[129] — Aggiunta alla Memoria "Preliminari di una teoria delle varietà luoghi di spazi". Rend. Circ. Mat. Palermo **30**, 346—348 $(1910)_2$.

[130] — Sugli elementi curvilinei che hanno comuni la tangente e il piano osculatore. Rend. Accad. Naz. Lincei (5) **33**, 325—329 $(1924)_1$.

[131] Severi, F.: Rappresentazione di una forma qualunque per combinazione lineare di più altre. Rend. Accad. Naz. Lincei (5) **11**, 105—113 $(1902)_1$; Memorie scelte. Vol. I, p. 445—454. Bologna: Zuffi 1950.

[132] — Sulle corrispondenze fra i punti di una curva algebrica e sopra certe classi di superficie. Mem. Accad. Sci. Torino (2) **54**, 1—49 (1903).

[133] — Su alcune questioni di postulazione. Rend. Circ Mat. Palermo **17**, 73—103 (1903).

[134] — Sulle relazioni che legano i caratteri invarianti di due superficie in corrispondenza algebrica. Rend. Ist. Lombardo (2) **36**, 499—511 (1903).

[135] — Su alcune proprietà dei moduli di forme algebriche. Atti Accad. Sci. Torino **41**, 167—185 (1905/06).

[136] — Sul metodo di Mayer per l'integrazione delle equazioni lineari ai differenziali totali. Atti Ist. Veneto **69**, 419—425 (1910).

[137] — Vorlesungen über algebraische Geometrie. Leipzig: B. G. Teubner 1921.

[138] — Trattato di geometria algebrica. Vol. I, parte 1ª. Bologna: Zanichelli 1926.

[139] — Nuovi contributi alla teoria delle serie di equivalenza sulle superficie e dei sistemi di equivalenza sulle varietà algebriche. Mem. Accad. Italia (1) **4**, 71—129 (1933).

[140] — Il rango di una corrispondenza a valenza sopra una superficie. Boll. Un. Mat. Ital. (1) **15**, 161—169 (1936).

[141] — La teoria generale delle corrispondenze tra varietà algebriche (2 Note). Rend. Accad. Naz. Lincei (6) **23**, 818—823; 921—925 $(1936)_1$.

[142] — La teoria generale delle corrispondenze fra due varietà algebriche e i sistemi d' equivalenza. Abh. Math. Semin. Hansischen Univ. **13**, 101—112 (1939).

[143] — Problema n. 118. Rend. di Mat. (5) **3**, 64—65 (1942).

[144] — Serie, sistemi d'equivalenza e corrispondenze algebriche sulle varietà algebriche (a cura di F. Conforto e E. Martinelli). Vol. I. Roma: Perrella 1942.

[145] — Les images géométriques des idéaux de polynomes. C. r. Acad. Sci. (Paris) **232**, 2395—2396 (1951).

[146] — Le diverse concezioni di varietà nella geometria algebrica. Rend. Accad. Naz. dei XL, (4) **2**, 155—181 (1951).

[147] Smith, P. A.: The topology of involutions. Proc. Nat. Acad. Sci. **19**, 612—618 (1933).

[148] — Fixed points of periodic transformations. Appendix B of S. Lefschetz, Algebraic topology (Amer. Math. Soc. Coll. Publ. Vol. **27**. New York 1942). p. 350—373.

[149] Sobrero, L.: Un teorema relativo ai sistemi differenziali. Rend. Accad. Naz. Lincei (6) **21**, 317—325 $(1935)_1$.

[150] — Delle funzioni analoghe al potenziale intervenienti nella Fisica Matematica. Rend. Accad. Naz. Lincei (6) **21**, 448—454 $(1935)_1$.

[151] Study, E.: Betrachtungen über Doppelverhältnisse. Leipziger Ber. **1896**, 200—220.

[152] Su, B.: Note on a theorem of B. Segre. Sci. Rep. Ac. Sinica **1**, 16—19 (1942).

[153] — Su alcuni invarianti di contatto di due varietà in uno spazio proiettivo. Boll. Un. Mat. Ital. (3) **1**, 9—12 (1947).

[154] Tafani, G.: Sulle corrispondenze $(1, n)$ tra varietà a 3 dimensioni. Ann. Sc. Norm. sup. Pisa (1) **13**, 3—44 (1919).

[155] Terracini, A.: Sulle linee proiettive di una superficie. Rend. Accad. Naz. Lincei (6) **22**, 125—129 $(1935)_2$.

[156] — Densità di una corrispondenza di tipo dualistico ed estensione dell'invariante di Mehmke-Segre. Atti Accad. Sci. Torino **71**, 310—328 (1935).

[157] — Invariante di Mehmke-Segre generalizzato e applicazione alle congruenze di rette. Boll. Un. Mat. Ital. (1) **15**, 109—113 (1936).

[158] — El invariante de Mehmke-Segre y los sistemas lineales, Anales Soc. Cient. Argentina **129**, 97—111 (1940).

[159] Tibiletti, C.: Gruppo concentrato di intersezioni di due curve algebriche. Rend. Ist. Lombardo **84**, 29—47 (1951).

[160] Torelli, R.: Sopra certe estensioni del teorema di Nöther $Af + B\varphi$. Atti Accad. Sci. Torino **41**, 186—196 (1905/06).

[161] Tricomi, F.: Sulla distribuzione dei baricentri delle sezioni piane di un corpo (2 Note). Rend. Accad. Naz. Lincei (6) **13**, 407—411; 478—484 $(1931)_1$.

[162] — "Densità" di un continuo di punti o di rette e "densità" di una corrispondenza. Rend. Accad. Naz. Lincei (6) **23**, 313—316 $(1936)_1$.

[163] Tzitzëica, G.: Géométrie différentialle projective des réseaux. Bucarest: Cultura Nationala 1923.

[164] Vahlen, K. Th.: Über den Grad der Eliminationsresultante eines Gleichungssystemes. J. reine u. angew. Math. **113**, 348—352 (1894).

[165] Waerden, B. L. van der: Algebra. II. Teil, 3. Aufl. der „Moderne Algebra", Berlin: Springer-Verlag 1955.

[166] Vesentini, E.: Sulle omografie definite da certe coppie di elementi differenziali tangenti. La Ricerca **3**, 26—28 (1952).

[167] Waelsch, E.: Zur Infinitesimalgeometrie der Strahlencongruenzen und Flächen. Sitzgsber. Wiener Akad. Wiss. **100**, 158—183 (1891).

[168] — Sur le premier invariant différentiel projectif des congruences rectilignes. C. r. Acad. Sci. (Paris) **118**, 736—738 (1897).

[169] Wedderburn, J. H. M.: Lectures on matrices. Amer. Math. Soc. Coll. Publ. Vol. **17**. New York 1934.

[170] Weiss, E. A.: Einführung in die Liniengeometrie und Kinematik. Leipzig: B. G. Teubner 1935.

[171] Wilczynski, E. J.: Projective differential geometry of curves and ruled surfaces. Leipzig: B. G. Teubner 1906.

[172] Zeuthen, H. G.: Nouvelle démonstration de théorèmes des séries de points correspondants sur deux courbes. Math. Ann. **3**, 150—156 (1871).

Author Index

Analytic Index

Ergebnisse der Mathematik und ihrer Grenzgebiete

1. Bachmann: Transfinite Zahlen. DM 42,—; US $ 11.60
2. Miranda: Partial Differential Equations of Elliptic Type. DM 58,—; US $ 16.00
4. Samuel: Méthodes d'Algébre Abstraite en Géométrie Algébrique. DM 29,—; US $ 8.00
5. Dieudonné: La Géométrie des Groupes Classiques. DM 42,—; US $ 11.60
6. Roth: Algebraic Threefolds with Special Regard to Problems of Rationality. DM 22,—; US $ 6.10
7. Ostmann: Additive Zahlentheorie. 1. Teil: Allgemeine Untersuchungen. DM 42,—; US $ 11.60
8. Wittich: Neuere Untersuchungen über eindeutige analytische Funktionen. DM 31,—; US $ 8.60
11. Ostmann: Additive Zahlentheorie. 2. Teil: Spezielle Zahlenmengen. DM 28,—; US $ 7.70
13. Segre: Some Properties of Differentiable Varieties and Transformations. DM 46,—; US $ 12.70
14. Coxeter/Moser: Generators and Relations for Discrete Groups. DM 35,—; US $ 9.70
15. Zeller/Beckmann: Theorie der Limitierungsverfahren. DM 64,—; US $ 17.60
16. Cesari: Asymptotic Behavior and Stability Problems in Ordinary Differential Equations. 3rd edition in preparation
17. Severi: Il teorema di Riemann-Roch per curve-superficie e varietà questioni collegate. DM 26,—; US $ 7.20
18. Jenkins: Univalent Functions and Conformal Mapping. DM 37,—; US $ 10.20
19. Boas/Buck: Polynomial Expansions of Analytic Functions. DM 18,—; US $ 5.00
20. Bruck: A Survey of Binary Systems. DM 46,—; US $ 12.70
21. Day: Normed Linear Spaces. DM 17,80; US $ 4.90
23. Bergmann: Integral Operators in the Theory of Linear Partial Differential Equations. DM 40,—; US $ 11.00
25. Sikorski: Boolean Algebras. DM 42,—; US $ 11.60
26. Künzi: Quasikonforme Abbildungen. DM 43,—; US $ 11.90
27. Schatten: Norm Ideals of Completely Continous Operators. DM 26,—; US $ 7.20
28. Noshiro: Cluster Sets. DM 40,—; US $ 11.00
29. Jacobs: Neuere Methoden und Ergebnisse der Ergodentheorie. DM 55,—; US $ 15.20
30. Beckenbach/Bellman: Inequalities. DM 38,—; US $ 10.50
31. Wolfowitz: Coding Theorems of Information Theory. DM 30,—; US $ 8.30
32. Constantinescu/Cornea: Ideale Ränder Riemannscher Flächen. DM 75,—; US $ 20.70
33. Conner/Floyd: Differentiable Periodic Maps. DM 29,—; US $ 8.00
34. Mumford: Geometric Invariant Theory. DM 24,—; US $ 6.60
35. Gabriel/Zisman: Calculus of Fractions and Homotopy Theory. DM 42,—; US $ 11.60
36. Putnam: Commutation Properties of Hilbert Space Operators and Related Topics. DM 31,—; US $ 8.60
37. Neumann: Varieties of Groups. DM 51,—; US $ 14.10
38. Boas: Integrability Theorems for Trigonometric Transforms. DM 20,—; US $ 5.50
39. Sz.-Nagy: Spektraldarstellung linearer Transformationen des Hilbertschen Raumes. DM 20,—; US $ 5.50
40. Seligman: Modular Lie Algebras. DM 43,—; US $ 11.90
41. Deuring: Algebren. DM 26,—; US $ 7.20
42. Schütte: Vollständige Systeme modaler und intuitionistischer Logik. DM 26,— US $ 7.20

43. Smullyan: First-Order Logic. DM 36,—; US $ 9.90
44. Dembowski: Finite Geometries. DM 68,—; US $ 17.00
45. Linnik: Ergodic Properties of Algebraic Fields. DM 44,—; US $ 12.10
46. Krull: Idealtheorie. DM 28,—; US $ 7.70
47. Nachbin: Topology on Spaces of Holomorphic Mappings. DM 18,—; US $ 5.00
48. A. Ionescu Tulcea/C. Ionescu Tulcea: Topics in the Theory of Lifting. DM 36,—; US $ 9.90
49. Hayes/Pauc: Derivation and Martingales. DM 48,—; US $ 13.20
50. Kahane: Séries de Fourier Absolument Convergents. DM 44,—; US $ 12.10
51. Behnke/Thullen: Theorie der Funktionen mehrerer komplexer Veränderlichen. DM 48,—; US $ 13.20
52. Wilf: Finite Sections of Some Classical Inequalities. DM 28,—; US $ 7.70
53. Ramis: Sous-ensembles analytiques d'une variété banachique complexe. DM 36,—; US $ 9.90
54. Busemann: Recent Synthetic Differential Geometry. DM 32,—; US $ 8.80
55. Walter: Differential and Integral Inequalities. DM 74,—; US $ 20.40
56. Monna: Analyse non-archimédienne. DM 38,—; US $ 10.50
57. Alfsen: Compact Convex Sets and Boundary Integrals. In preparation
58. Greco/Salmon: Topics in $\mathfrak{m}$-Adic Topologies. DM 24,—; US $ 6.60
59. López de Medrano: Involutions on Manifolds. DM 36,—; US $ 9.90
60. Sakai: C*-Algebras and W*-Algebras. In preparation